A MATHEMATICAL TAPESTRY

Demonstrating the Beautiful Unity of Mathematics

This easy-to-read book demonstrates how a simple geometric idea reveals fascinating connections and results in number theory, polyhedral geometry, combinatorial geometry, and group theory. Using a systematic paper-folding procedure, it is possible to construct a regular polygon with any number of sides. This remarkable algorithm has led to interesting proofs of certain results in number theory, has been used to answer combinatorial questions involving partitions of space, and has enabled the authors to obtain the formula for the volume of a regular tetrahedron in around three steps, using nothing more complicated than basic arithmetic and the most elementary plane geometry. All of these ideas, and more, reveal the beauty of mathematics and the interconnectedness of its various branches.

Detailed instructions, including clear illustrations, enable the reader to gain hands-on experience constructing these models and to discover for themselves the patterns and relationships they unearth.

PETER HILTON is Distinguished Professor Emeritus in the Department of Mathematical Sciences at the State University of New York (SUNY), Binghamton.

JEAN PEDERSEN is Professor of Mathematics and Computer Science at Santa Clara University, California.

SYLVIE DONMOYER is a professional artist and freelance illustrator (www.scientific-illustrator.com).

A MATHEMATICAL TAPESTRY

Demonstrating the Beautiful Unity of Mathematics

PETER HILTON
State University of New York, Binghamton

JEAN PEDERSEN
Santa Clara University, California

With illustrations by

SYLVIE DONMOYER

CAMBRIDGE
UNIVERSITY PRESS

CAMBRIDGE UNIVERSITY PRESS
Cambridge, New York, Melbourne, Madrid, Cape Town, Singapore,
São Paulo, Delhi, Dubai, Tokyo, Mexico City

Cambridge University Press
The Edinburgh Building, Cambridge CB2 8RU, UK

Published in the United States of America by Cambridge University Press, New York

www.cambridge.org
Information on this title: www.cambridge.org/9780521764100

First published 2010
Reprinted 2010

Printed in the United Kingdom at the University Press, Cambridge

A catalogue record for this publication is available from the British Library

Library of Congress Cataloguing in Publication data
Hilton, Peter John.
A mathematical tapestry : demonstrating the beautiful unity of mathematics / Peter Hilton, Jean Pedersen ;
with illustrations by Sylvie Donmoyer.
p.   cm.
Includes bibliographical references and index.
ISBN 978-0-521-76410-0 (hardback)
1. Mathematics.   I. Pedersen, Jean.   II. Title.
QA36.H53   2010
510 – dc22   2010010230

ISBN 978-0-521-76410-0 Hardback
ISBN 978-0-521-12821-6 Paperback

This book is dedicated
to the memory of
Martin Gardner
(1914–2010)

Contents

Preface

This is a book of 17 chapters, each of which provides some arithmetic, some geometry, or some algebra. The basic ideas in each chapter we call ***threads***, and there are at least nine threads in this book – paper-folding threads, number-theory threads, polyhedral threads, geometry threads, algebra threads, combinatorial threads, symmetry threads, group-theory threads, and historical threads. So this book utilizes, exploits, and develops, by weaving these threads of a very different kind together, many parts of mathematics. At the end of this preface we will give a table showing how you might read this book in the very unlikely event that you are interested in just one of these threads.

Many of the chapters involve the construction of models and these take time and effort, but we believe that if you choose to carry out the constructions you will find the activity satisfying. As Benno Artmann, reviewing one of Pedersen's articles in *Mathematical Reviews*, said about the construction of the golden dodecahedron: "I tried a dodecahedron. It sits on my desk, looks nice and makes me feel like an artist." On the other hand, we understand that many of our readers won't want to construct these models, but we think that they can be appreciated without actually constructing them. Surely you yourselves have enjoyed eating a piece of cake that someone else baked; and, even though you didn't actually do the baking, you might be interested in its ingredients and how it came to be in its final shape. So we include the instructions for building the models for those of you who want to construct them and hope that our other readers will at least appreciate what goes into making them.

We have had the immense benefit of the cooperation of the artist Sylvie Donmoyer who has provided beautiful, highly illustrative pictures, Hans Walser who has provided the figures for Chapter 13, and Byron Walden who took responsibility for the proofs in Chapter 17.

In addition to the various threads, there is one technical feature of this book which we would like to mention. We adopt, when appropriate, the notation of using radian measure, writing π (without the word "radians" following it) instead of $180°$. The advantages of adopting this notation will, we believe, become obvious to our readers when it is used; since it clearly emphasizes that the straight angles of $180°(= \pi)$ at the edge of the strip of paper being folded play a special role in the geometry, and subsequently in the number theory – as they obviously do. A feature which we use throughout the book, to make it easier for the reader to spot when a new idea is being named, is to use **bold italic** print to alert you that we are introducing a technical term that will be used subsequently. On the other hand, we use *ordinary italic* print for emphasis. We – and our readers – have found this convention helpful in some of our previous publications.

The topics of this book were chosen because of their interconnectedness; and the aim of the book is to show how pursuing a single idea in mathematics can lead in many different directions forming a unified whole. We think this book should be of interest to bright high-school students and all other intelligent people with an interest in mathematics because it touches on so many fascinating aspects of mathematics and the people who do it. We give some highlights below for number-theorists and geometers.

For those interested in number theory there are at least two very significant theorems in this book. It is intriguing to think that these two striking theorems about numbers came about naturally by following the mathematics of paper-folding described in Chapter 2, which was motivated by the hexaflexagons introduced in Chapter 1. Just to whet your appetite we will tell you now that the first of these important theorems, the quasi-order theorem, enables one to determine for any given odd number $b \geq 3$, using an algorithm that involves only subtraction and division by the number 2, the smallest power k to which 2 must be raised in order that either $2^k - 1$, or $2^k + 1$ is exactly divisible by b. And it tells us whether the sign should be "−" or "+". Furthermore, the algorithm never uses any number larger than b itself. The proof of this theorem is the most delicate result we present, but it seems, in the context of our development, to be a perfectly natural result. It leads, indeed, to a proof that the Fermat number, F_5, which is $2^{32} + 1$, is not prime (see Chapter 7). It is, in fact, the smallest Fermat number to be composite. Our proof is based on an algorithm that uses only subtraction and division by 2, and involves no number larger than 641.

The second significant number-theoretic result occurs in Chapter 7. We call that result the coach theorem because the mathematical symbols in the statement of the theorem look like coaches on a train to the English author, and we yielded to his wording, since to have called it a "car theorem" (because Americans refer to cars on a train) didn't sound nearly so nice to either of the authors. In Chapter 17 the coach theorem enables us, by a logical extension of the quasi-order theorem, to determine for any given $b \geq 3$ the number of proper divisors of the number b. In other words it gives us the value of what is well-known among number-theorists as the Euler totient function of b, that we denote $\Phi(b)$ (as defined in Section 17.1). This result is obtained through repeated use of the algorithm used for the quasi-order theorem. We have described both of these theorems in terms of the number 2 but they both have generalizations involving a general positive number $t \geq 2$, which we also give in Chapter 17 for the benefit of those truly interested in number theory.

There are also results about divisibility that are quite counter-intuitive. For example, in Chapter 4 we take the basic number fact $7 \times 3 = 21$ and show that the two related number facts

$$7 \text{ divides } 21$$

and

$$3 \text{ divides } 21$$

have very different generalizations to arbitrary bases!

Readers who are especially interested in geometry will find here a completely systematic method for constructing arbitrarily good approximations to regular b-gons, for *any* $b \geq 3$; this is a result that we believe the Greeks and Gauss would surely have liked to know about. These constructions led one of the authors to discover a construction of braided polyhedra (Chapters 8 and 9) with remarkable geometric properties of their own. Surprisingly, these same braided polyhedra are useful in determining the number of unbounded regions created in space by the extended face planes of the Platonic solids (Chapter 14).

The following table suggests which chapters to read for those of you with an overpowering interest in one or another of the nine threads. It should not be assumed, however, that if a chapter is not mentioned in one of the threads, then the thread does not appear at all there – this table lists the chapters that have some substantial parts devoted to the topic mentioned. For example, you may note that all chapters are listed under symmetry; and you will find that some are of an intensely geometric nature, while in other chapters the symmetry is in the statement of number-theoretic results, and in yet other chapters the symmetry is only present subtly without being mentioned explicitly.

Thread	Chapters
paper-folding	2, 3, 7
number theory	2, 3, 4, 7, 17
polyhedral	5, 6, 8, 9, 10, 11, 12, 13, 14, 15
geometry	1, 2, 3, 5, 6, 8, 9, 10, 11, 12, 14, 15, 16
algebra	1, 2, 3, 4, 5, 7, 17
combinatorial	10, 12, 14
symmetry	1–17
group theory	9, 11, 13, 14, 16
historical	1, 10, 15, 16

We realize that not all readers will be interested in reading highly technical proofs. So we have placed an asterisk by the titles of certain sections, where the mathematics gets more intense, to let you know you can, with impunity, skip all (or part of) those sections on first reading and go straight on to the examples in those sections, or to other concepts. In almost all cases you will still be able to understand the subsequent material without having digested the proofs.

Acknowledgments

We are grateful to Don Albers for his support and guidance during the pre-publication stage of this book, and to Silvia Barbina, Clare Dennison, and Dawn Preston at Cambridge University Press, along with the copy-editor, Mike Nugent, who shepherded this project through the final publishing process. Alas! Long before the book came to be, we were encouraged by, and received excellent advice from, our colleagues and students about the content of the book – and we owe all of them a debt of gratitude for their enthusiasm and helpfulness.

We particularly wish to thank Gerald L. Alexanderson, Monika Caradonna, Victor Garcia, Jennifer Hooper, and Byron Walden for their careful attention to detail in helping with the proof-reading of this book. We also owe a huge special thanks to our spouses, Margaret Hilton and Kent Pedersen, and to our families for their ongoing support, and encouragement, of our joint work over the past 30 years!

Peter Hilton   Binghamton, New York
Jean Pedersen   Santa Clara, California
                                   April 2010

1

Flexagons – A beginning thread

1.1 Four scientists at play

In 1939 four young men were thrown together in graduate school at Princeton. They came from diverse backgrounds and went on to have very different careers. But for a short while all of them played with straight strips of paper made into what became known as "flexagons." The thread that runs through this story is that of four creative young men who played for a while with an intriguing toy and then went on to other creative ventures in mathematics, physics, statistics, and computer science. We present here a small bit from each of their life stories.

In 1939 Arthur H. Stone (1916–2000), then a newly transplanted Englishman, was beginning his PhD work with Solomon Lefschetz at Princeton. According to Paul M. Cohn's obituary of Stone [6],

Arthur Harold Stone . . . was one of the foremost general topologists of his time, and made significant contributions to a number of different parts of general topology. . . . In 1927 [he] won a LCC scholarship to Christ's Hospital (Horsham). This was a boarding school which has had such successful pupils as Philip Hall, Christopher Zeeman (later Sir Christopher) and D. G. Northcott (Stone's contemporary). The mathematics teaching was in the hands of C. A. J. Trimble, himself a Wrangler. Here, Arthur won prizes in almost all subjects except sports (though he was also good at rugger [rugby football]).

In 1935 he gained a major scholarship to Trinity College, Cambridge. He excelled at the academic subjects, but was also an outstanding violinist and good at chess. At Cambridge he continued with the violin and became leader of the orchestra of the Cambridge University Music Society. He was a Wrangler, and took his BA in 1938 before going to Princeton to work for a PhD with S. Lefschetz.

To fit the American notebook sheets into his English binder, he had to trim off an inch of paper, and he began to fold these strips in various ways. This led to some intriguing figures, which later became famous as 'flexagons' (see [17]). He was both very inventive and also adept with his hands, talents which he used in building a counterclockwise grandfather clock.

The popularization of flexagons began when Martin Gardner wrote about them in his first *Scientific American* article [16]; and it was that article that led to his regular *Scientific American* feature article on Mathematical Games. Gardner's article was closely followed by Oakley and Wisner's more mathematical paper on flexagons [54].

The quote in the first obituary above, from [17], gives a fascinating account of Stone's discovery and his collaboration with Bryant Tuckerman (1915–2002), Richard P. Feynman (1919–1988), and John W. Tukey (1915–2000). Each of these collaborators became a famous scientist in his own right – but none of their subsequent successes seems to have been directly concerned with flexagons.

Bryant Tuckerman's obituary [75] states that

His graduate studies were interrupted by World War II, during which he worked at the U. S. Office of Scientific Research and [the] Office of Scientific Research and Development on navigational devices for tanks. After the war he completed his PhD at Princeton in topology. . . . He then worked for five years at the Institute for Advanced Study in Princeton with John von Neumann on applications of such early computers as the MANIAC. . . . In 1962 Tuckerman published *Planetary, Lunar, and Solar Positions*, a set of tables covering the years from 601 B.C. to 1649, which is still used by historians and archaeologists to date ancient documents containing astronomical references, from Babylonian times through the Renaissance. In 1971 he found the 24th Mersenne prime, $2^{199937} - 1$, then the largest known prime number.

Richard Feynman was awarded the Nobel Prize in physics in 1965, along with Shinichero Tomonaga of Japan and Julian Schwinger of Harvard University. On his death it was reported by Chandler in Feynman's obituary [5] that

He was widely known for his insatiable curiosity, gentle wit, brilliant mind and playful temperament. These qualities were clearly evident in his popular 1985 book, "Surely You're Joking Mr. Feynman," which was on the New York Times best-seller list for 14 weeks. . . . Ever playful and unintimidated by authority, Mr. Feynman caused consternation in his years with the Manhattan Project, which developed the atomic bomb, by figuring out in his spare time how to pick the locks on filing cabinets that contained classified information. Without removing anything, he left taunting notes to let officials know that their security system had been breached. . . .

Mr. Feynman attracted widespread attention during the Rogers Commission hearings on the Challenger space shuttle accident in 1986. Frustrated by witnesses' vague answers and by slow bureaucratic procedures, he conducted an impromptu experiment that proved a key point in the investigation: He dunked a piece of the rocket booster's O-ring material into a cup of icewater and quickly showed that it lost all resiliency at low temperatures. . . .

MIT physicist Philip Morrison called Mr. Feynman "the most original theoretical physicist of our time," according to a report by United Press International. Morrison said Mr. Feynman, who called his Nobel Prize "a pain in the neck" was "extraordinarily honest with himself and everyone else," and added that "he didn't like ceremony or pomposity . . . he was extremely informal. He liked colorful language and jokes."

As for John Tukey, *The New York Times* reported in 2000:

In 1936, Mr. Tukey graduated from nearby Brown University with a bachelor's degree in chemistry, and in the next three years earned three graduate degrees, one in chemistry at Brown, and two in mathematics at Princeton, where he would spend the rest of his career. At the age of 35 he became a full professor, and in 1965 he became the founding chairman of Princeton's statistics department.

[He was known as] one of the most influential statisticians of the last 50 years and a wide-ranging thinker. . . . Three decades before the founding of Microsoft, Mr. Tukey saw that "software" as he called it, was gaining prominence. "Today", he wrote at the time it is "at least as important" as the "hardware" of tubes, transistors, wires, tapes and the like. . . . Twelve years earlier, while working at Bell Laboratories he had coined the term "bit," an abbreviation of "binary digit" that described the 1's and 0's that are the basis of computer programs. Both words caught on, to the chagrin of some computer scientists who saw Mr. Tukey as an outsider. "Not everyone was happy that he was naming things in their field," said Steven M. Schultz, a spokesman for Princeton. . . .

As his career progressed, he also became a hub for other scientists. He was part of a group of Princeton professors that gathered regularly and included Lyman Spitzer, Jr., who inspired the Hubble Space Telescope.

His first brush with publicity came in 1950, when the National Research Council appointed him to a committee to evaluate the Kinsey Report, which shocked many Americans by describing the country's sexual habits as far more diverse than had been thought. From their first meeting, when Mr. Kinsey told Mr. Tukey to stop singing a Gilbert and Sullivan tune aloud while working, the two men clashed, according to "Alfred C. Kinsey", a biography by James H. Jones.

In a series of meetings over two years Mr. Kinsey vigorously defended his work which Mr. Tukey believed was seriously flawed, relying on a sample of people who knew each other. Mr. Tukey said a random selection of three people would have been better than a group of 300 chosen by Mr. Kinsey.

Nothing that any of these four brilliant men did would have led one to suppose that they would jointly invent a very special brand of polygon, namely, a flexagon. It is interesting to point out that, of the four pioneers of flexagons, Arthur Stone was the only professional mathematician.

1.2 What are flexagons?

In the next section we will describe in detail how to construct flexagons; that is, special polygons that change their appearances when they are manipulated in certain ways. But first we need some terminology.

In general, we refer to these configurations as *N-flexagons*, where N indicates the number of congruent triangles surrounding the center of the regular polygon formed by the completed construction. We should point out that in the case of the 8-flexagon, the bounding polygon of the construction has 4 sides, not 8 as you might expect. For this reason the 8-flexagon is often referred to in the literature as a *tetraflexagon*.

We first describe in detail how to construct and flex two special, different types of flexagon, for each of the values $N = 6$ and $N = 8$. We also refine the nomenclature by adding another number at the beginning of the name to indicate how many different complete "faces," of N triangles each, the model may present when it is flexed. This nomenclature will be illustrated and described in detail at the appropriate time – and, for those of you who like to use big words, we will also give you the names that utilize Greek prefixes instead of numbers.

We first discuss, in Section 1.3, the hexaflexagons (which we call **6-*flexagons***). We do this because they are the easiest flexagons to manipulate. Our idea is that you are likely to find it more pleasant to develop your flexing skills on the 6-flexagons, and that by doing this you will be better prepared to appreciate the considerably more complicated octaflexagons (or 8-flexagons) that follow in Section 1.4. Of course, in keeping with the spirit of this entire book, we suggest variations or references along the way so that you can construct (or maybe even *invent*) other flexagons on your own.

1.3 Hexaflexagons

Required materials

- Strips (or a roll) of gummed mailing tape or adding machine tape about $1\frac{1}{2}$ inches (4 cm) wide. The glue on the gummed tape should be of the type that needs to be moistened to become sticky.
- White glue, or a glue stick, if your folding tape is not gummed.

The simplest hexaflexagon is made from a straight strip of 10 equilateral triangles (if you want an easy way to construct these triangles see Section 2.3). We describe the construction here in terms of instructions of a kind that we will use to show how to make any of our flexagons.

Basic instructions

1. Prepare the pattern piece, labeling both sides of it *precisely* as shown.
2. Crease all fold lines in *both* directions.
3. Beginning with the strip as shown at the top of Figure 1.1, fold *in order* (so that the numbers are no longer visible),

    triangle 1 onto triangle 1,
    triangle 2 onto triangle 2,
    triangle 3 onto triangle 3,

    •
    •
    •

    and, finally, triangle ☆ onto triangle ☆.

4. Glue, or attach with a paper clip, so that ☆ is attached to ☆.
5. Gently flex and play with your model – decorate the faces with interesting patterns.

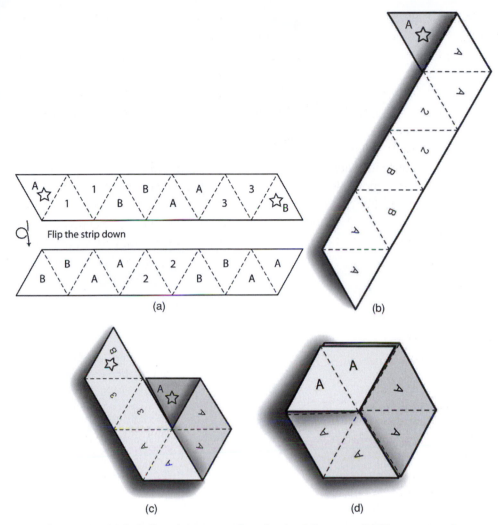

Figure 1.1   (a) Labeling the pattern piece for the 6-flexagon. (b) The construction with triangle 1 on triangle 1, (c) with triangle 2 on triangle 2, (d) with triangles 3 on triangle 3 and nearly completed.

Figure 1.1(a) shows the pattern piece for the 6-flexagon. A word of caution: Note that when you "flip the strip *down*," as indicated by the symbol between the two strips, you simply take the top edge and lift it off the table and place it at the bottom (nothing moves in the right- or left-hand direction). Place the strip on the table with the side showing the numbers 1 and 3 visible. Then follow the instructions. Figures 1.1(b)–(d) should reassure you that you are doing it correctly. Try out the construction.

Now for the magic! Gently mountain-fold and valley-fold the hexagon as shown in Figure 1.2(a) to make a 3-petaled arrangement that will "come apart" at the

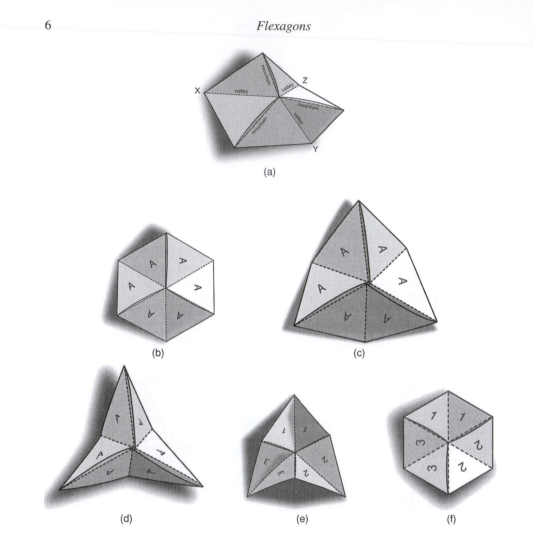

Figure 1.2   (a) Beginning to flex. Note the location of the "slits" on the mountain folds. (b)–(f) How the flex proceeds, showing the labeled faces.

top and lie flat when the vertices labeled *x*, *y*, *z* are brought together *below* the hexagon. Following the illustrations in Figures 1.2(b)–(f) may be helpful. Repeat the process. Notice that as you flex the hexagon in this way you eventually see 3 faces, the A face, the B face, and the 1-2-3 face.

Although 6-flexagons constructed from the same width tape will all have the same shape and size, they may differ in the number of hexagonal faces that can be presented as the polygon is flexed. In this sense, the flexagon you have just constructed is the "smallest" hexaflexagon that can be constructed with a *straight* strip of equilateral triangles. Since it has 3 faces, we call it a 3-6-flexagon (it is well-known by the name of ***trihexaflexagon***).

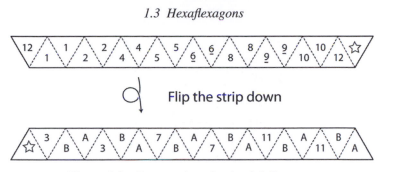

Figure 1.3   Pattern piece for the 6-6-flexagon.

Play with your flexagon until you become adept at flexing it. You may want to draw some patterns on its faces. Begin by drawing a design on the two visible faces and then flex it. You will notice a blank face appears and one of the existing faces disappears – and even the face that is still visible may seem changed. You will soon see that although you can draw patterns on only 3 hexagonal faces originally, more than 3 designs will appear, owing to the way the patterns on the triangular portions of the face are moved about when the flexagon is flexed.

A 6-6-flexagon (the **hexahexaflexagon**) may be constructed from a strip of 19 equilateral triangles. The pattern piece is shown in Figure 1.3. Now it's up to you! Following the basic instructions, make your 6-6-flexagon and then read the rest of this section for suggestions about how to flex it and how to build even bigger 9-6-flexagons, 12-6-flexagons, and, in general, $3n$-6-flexagons. *A practical hint*: As you complete each "triangle" so that an A is on one side and a B is on the other side, place a paper clip on that triangle and continue to fold the next triangle, placing like numbers on each other. When you have completed the flexagon remove all the paper clips. At that point there should be 6 triangles labeled A on one face and 6 triangles labeled B on the other face.

The 6-6-flexagon is flexed in exactly the same way as the 3-6-flexagon. However, most people have difficulty finding all 6 faces. Bryant Tuckerman invented a procedure for bringing out the 6 faces with the shortest possible flexing sequence. His process, known as the ***Tuckerman traverse***, involves continually flexing at one vertex until the flexagon refuses to open, then moving to an adjacent vertex (either way) and continuing to flex at that vertex until the flexagon again refuses to open, and so on. It is an interesting exercise to record the *sequence of faces* that appears as you perform this flexing algorithm. Try it and compare your results with the diagram in Figure 1.4.

Notice that although you drew 6 patterns on the faces of this polygon, there are many more actual designs (since the triangular parts of the hexagon appear in different orientations as you flex the model). How many different designs do you

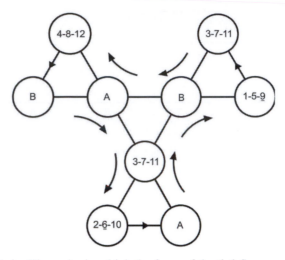

Figure 1.4   The order in which the faces of the 6-6-flexagon appear.

get from your 6 faces? (Your answer will depend on the symmetry of the patterns you use.)

You may have already noticed a pattern in the *number of faces* on these 6-flexagons. The smaller had 3 faces, and the last one had 6 faces. It is a fact (which we don't prove) that the number of faces that can occur on a 6-flexagon constructed with straight strips of equilateral triangles must be a multiple of 3. So, of course, the next larger 6-flexagon will have 9 faces. But, how do we construct it? Here are some hints.

First of all let us suppose that somehow we remember that it is possible to construct a 3-6-flexagon but that we've forgotten how many triangles we need. We can readily calculate the number of triangles required. We need to have 6×3 triangles available in order to provide 3 faces. We also need 2 extra triangles that get glued together. Thus this flexagon contains $6 \times 3 + 2$ triangles in all. However, since each triangle on the strip of tape has 2 sides, the number of triangles this model actually requires is only half this number, that is

$$\frac{\mathbf{3} \times 6 + 2}{2} = 3 \times 3 + 1 = 10.$$

In exactly the same way, we can reason that for a 6-faced 6-flexagon the number of triangles required is

$$\frac{\mathbf{6} \times 6 + 2}{2} = 3 \times 6 + 1 = 19,$$

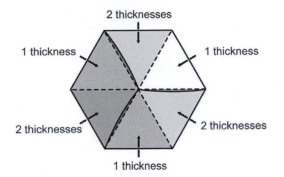

Figure 1.5 $1 + 2 = 3 =$ the number of faces for a 3-6-flexagon.

so that, for the **3n**-faced 6-flexagon the number of triangles required is

$$\frac{\mathbf{3n} \times 6 + 2}{2} = 3n \times 3 + 1 = 9n + 1.$$

Thus, for example, the 9-6-flexagon (*nonahexaflexagon*) requires a strip of $9 \times 3 + 1 = 28$ triangles.

Now that we know the number of triangles required for our 9-6-flexagon, how do we get them folded in the right arrangement? Again, we study the two 6-flexagons we've already constructed.

Notice that, in the flattened position of the 3-6-flexagon, the thicknesses of tape on two adjacent triangular sections are 1 and 2, respectively (see Figure 1.5). (Where two triangles are glued together, they behave as 1 thickness of tape.)

What is the situation with the 6-6-flexagon? We observe that in its flattened position immediately after construction (*before* any flexing takes place), the thicknesses of tape on two adjacent triangular sections are 2 and 4. However, when the 6-6-flexagon is flexed, it sometimes has thicknesses of 1 and 5 on adjacent triangular sections. See Figure 1.6.

This information contains the secret for constructing the 9-6-flexagon. What we might seek is an arrangement so that the thicknesses on any two adjacent triangular sections of the flattened hexagon sum to 9. One possibility is to use the fact that $4 + 5 = 9$ and try to find out how to fold the strip of equilateral triangles so as to produce adjacent triangles on the finished model having 4 and 5 thicknesses, respectively (see Figure 1.7). But we already know, from our construction of the 6-6-flexagon, how to fold the strip to obtain 4 thicknesses on one of the triangular sections; and, as we've observed, there must exist a way to obtain 5 thicknesses on a triangular section. The idea is to construct the 6-6-flexagon, except that you attach the two faces together with a paper clip instead of using glue. You can then flex this flexagon until you have thicknesses of 1 and 5 on adjacent triangular sections.

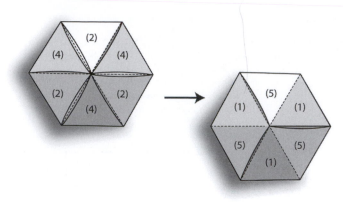

Figure 1.6 $2 + 4 = 1 + 5 = 6 =$ the number of faces for the 6-6-flexagon.

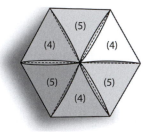

Figure 1.7   Thicknesses for the triangular sections of the 9-6-flexagon.

At that point you can remove the paper clip and "unwrap" the arrangement to *see* how to fold a triangular section with 5 thicknesses. With a little practice you will be surprised how easily you can guess how to fold the required number of thicknesses for a given triangular section.

In the same way that we figure out from the 6-6-flexagon how to construct the 9-6-flexagon with a strip of 28 equilateral triangles, we can use the 9-6-flexagon to discover how to build the 12-6-flexagon (*dodecahexaflexagon*) with a strip of 37 equilateral triangles. Of course, the process goes on, and you may even find that if you try it you can construct the 15-6-flexagon with a strip of 46 equilateral triangles. In fact, constructing it may be easier for some people than learning how to pronounce its Greek name, which is *pentacaidecahexaflexagon*! (*Pentacaideca* means "5 and 10" or "15" and, of course, "hexa" means "6".)

On a practical note we should warn you that if you construct flexagons with more than 6 faces it is best to trim a small amount of paper from each edge of the strip after folding the triangles before constructing the flexagon, so that the thickness of the paper doesn't get in the way of either the construction or the flexing of the finished product. We can't specify how much you should trim, because the

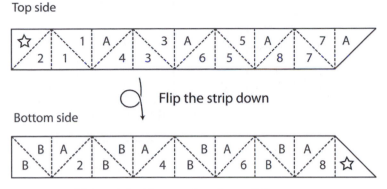

Figure 1.8   Pattern piece for an 8-flexagon.

thickness of the paper you use determines this, so experiment to find out how much is best for the paper you use. Some people even like to trim just a tiny amount from the edges when making the 6-6-flexagon, in order to make it flex more easily.

1.4 Octaflexagons

The octaflexagon (8-flexagon) was discovered independently by many people including one of the authors [61] and is often referred to as a tetraflexagon because the bounding polygon of the original construction is a square. However, we take a different point of view and call this an 8-flexagon because there are 8 hinged triangular sections surrounding the center of the polygon – just as there are 6 hinged triangular sections surrounding the center of the 6-flexagons.

There are many differences (and some similarities) between 6-flexagons and 8-flexagons. The first difference that should concern us is that 6-flexagons are constructed from straight strips of equilateral triangles, whereas 8-flexagons are constructed from straight strips of isosceles right triangles. Since these are exact constructions, involving the creation of angles of size $\frac{\pi}{2}$ and $\frac{\pi}{4}$, we are confident that our readers will be able to figure out for themselves how to make the pattern piece shown in Figure 1.8.

Once the pattern piece is folded, all you need to do to produce your 4-8-flexagon is follow the basic instructions in Section 1.3.

One word of caution before you proceed. This flexagon is much more versatile than the 6-flexagons. It can assume many shapes other than the square, and there are at least three different ways it can be flexed. As a result of the 8-flexagon's extraordinary capability to display different faces and shapes to the world, it sometimes gets twisted. This is not serious, and, if you are patient, it can always be untwisted. However, if you feel that it is possible that you might suffer from lack

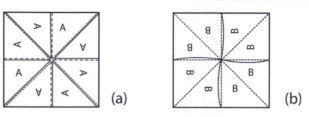

Figure 1.9   Two faces of the 8-flexagon.

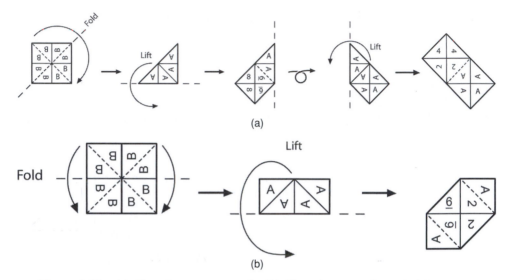

Figure 1.10   (a) Changing to a rectangle. (b) Changing to an (irregular) hexagon.

of patience, we offer this bit of advice: in the beginning, paper-clip the last 2 faces together instead of gluing them. This way, if you inadvertently get the flexagon in a state that frustrates you, then you can simply remove the paper clip and reassemble it. This procedure also allows you to unwrap the flexagon to find out whether or not you have put patterns on all the faces.

Now get, or construct, your 4-8-flexagon and then return. We'll wait.

Look at your flexagon. Observe that there are subtle differences between the 2 visible faces. One face should look like (a) and one should look like (b) in Figure 1.9.

Surprisingly, this flexagon can *change its shape*. To see how this happens, begin with the (a) side up to execute both of the moves shown in Figure 1.10. Of course, after you have gone from left to right you will need to reverse the moves to get the flexagon back into its original shape. You may wish to practice these procedures until you have a feel for them. Take your time. Then come back and we'll tell you how to flex your 8-flexagon in ways similar to your procedures with the 6-flexagon.

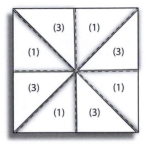

Figure 1.11   Thicknesses on the adjacent triangles of a 4-8-flexagon.

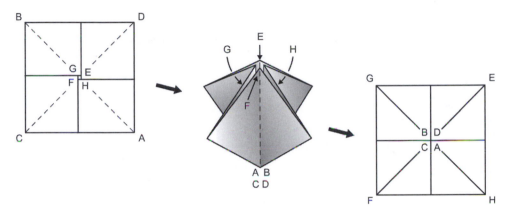

Figure 1.12   Executing the *straight flex*.

Before we begin flexing the 4-8-flexagon, let us see how we could anticipate how many faces it has. Note that the thicknesses on adjacent triangular sections are 1 and 3 (see Figure 1.11). Thus we might expect (and it turns out to be true) that this flexagon will have 4 distinct faces. We challenge you to find those 4 faces by repeated flexing.

We describe first a *straight flex*. Begin with your flexagon in the position shown at the start of Figure 1.12 (if it doesn't look exactly like this, turn it over). Then make valley folds along the dotted lines, and bring the 4 vertices labeled A, B, C, and D together below the flexagon to get the center figure. Then pull apart the top of the flexagon which will lie flat again in the shape of a square as shown in the last figure. To repeat this straight flex, as we call it, you must turn the flexagon over. Practice this a few times and draw patterns on all the faces you can find.

You are now ready for the more complicated *pass-through flex*. To do this, begin with the other side of the flexagon facing you (as shown at the start of Figure 1.13) and make mountain folds along the diagonal lines so that you obtain a 4-petaled arrangement. Then pull 2 opposite petals apart and down. You will then have a square platform above the 2 petals you pulled down. Fold the sides of the

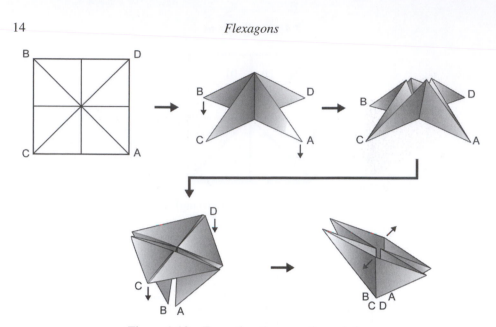

Figure 1.13   Executing the *pass-through flex*.

platform down and let the flexagon open at the top as shown in the last part of Figure 1.13. Be patient, keep calm, stay happy! It will work! (In the very unlikely event that you don't get it right the first time, try again – the very worst that could happen is that you will have to take your flexagon apart, reassemble it, and then try again.)

Practice flexing with the straight and pass-through flexes. How many faces can you find? (*Answer*: Only 3.) However, there is yet another way to flex. We call it the *reverse pass-through flex*, and we've saved it for last because it really is a little tricky. To perform this flex you do all the steps of the pass-through flex, but just at the point where you would open the flexagon, *STOP*! At this point your flexagon looks like 4 petals that are formed by 4 mountain folds and 4 valley folds. Now you complete the reverse pass-through flex by *reversing* the mountain and valley folds (as shown in Figure 1.14) and then opening the flexagon at the top as before. *Now* you will see the fourth face.

After you have become familiar with your 4-8-flexagon you will be ready for a real treat – the 8-8-flexagon (***octaoctaflexagon***). This model is no more complicated to construct than the 4-8-flexagon, and it is flexed in exactly the same ways.

We now have a special challenge for you. Recall that we know how to bring out all the faces on a 6-flexagon with a simple algorithm, called the ***Tuckerman traverse***. We therefore felt sure that there must be a fairly simple algorithm for bringing out all the faces of an 8-flexagon. In fact, such an algorithm was shown to us by JP's daughter, Jennifer Hooper (née Pedersen). But we should warn you that although each face appears on the top *or* bottom in the course of executing

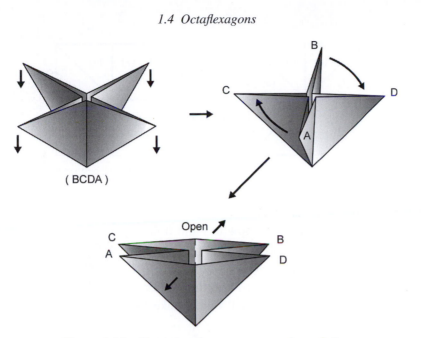

(BCDA)

Open ↗

Figure 1.14   Executing the *reverse pass-through flex*.

Top side

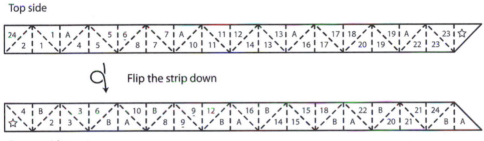

Flip the strip down

Bottom side

Figure 1.15   Pattern piece for the 8-8-flexagon.

this algorithmic sequence of flexes, it is *not* true that each face will appear on the top (as happens with the Tuckerman traverse for hexaflexagons) nor that each face will appear on the bottom. Can you find Jennifer's algorithm? We'll give you a hint. Her algorithm involves only the straight flex and the reverse-through flex. But this is not surprising, since, as you may verify, the pass-through flex has the same effect as the following sequence of three flexes:

reverse pass-through flex – straight flex – reverse pass-through flex.

As you construct the 8-8-flexagon from the pattern piece shown in Figure 1.15, you may notice that the thicknesses on adjacent triangular sections are 3 and 5 (which accounts for the 8 faces). Notice, however, as you flex your 8-8-flexagon, that it is possible to produce an arrangement where the thicknesses on adjacent

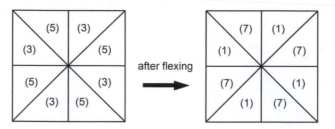

Figure 1.16   Thicknesses on adjacent faces of an 8-8-flexagon.

triangular sections will be 1 and 7 (see Figure 1.16). This is marvelous! It means that, just as we could figure out how to build 6-flexagons with $3n$ faces from 6-flexagons with fewer faces, we can figure out how to build 8-flexagons with $4n$ faces from 8-flexagons with fewer faces.

We now close this chapter and leave the further exploration of these ideas to you. We feel certain some of you will want to take a strip of

$$\frac{\mathbf{12} \times 8 + 2}{2} = 49$$

isosceles right triangles and build the **12**-8-flexagon (*dodecaoctaflexagon*), which has 5 and 7 thicknesses on adjacent triangular sections. You may even want to take a strip of

$$\frac{\mathbf{16} \times 8 + 2}{2} = 65$$

isosceles right triangles and build the **16**-8-flexagon. We could even believe that some people might want to make this giant flexagon just so that, when there is a lull in the conversation, they can talk about their *hexacaidecaoctaflexagon*!

There is much more to the story about flexagons (see, for example, [71]). A brief search of the Web for the word "flexagons" or "hexaflexagons" will turn up the most recent publications, and inventions, in this area. Not surprisingly, to people who see group theory everywhere, there is some group theory connected with some (but not all!) of the flexagons in this chapter. We discuss the group theory concerning the 3-6-flexagon in Chapter 13.

2

Another thread – 1-period paper-folding

2.1 Should you always follow instructions?

All self-respecting human beings, and therefore all our readers, must answer this question with a resounding NO! In the next paragraph we describe two aspects of our paper-folding, and building, instructions where we do advise rather rigid adherence to our specifications. However, we are very far from recommending that you fold all your regular polygons and construct all your polyhedra *exactly* as described. What we have done is to give you *algorithms* for the relevant constructions. Machines follow algorithms with relentless fervor, while human beings look for special ways of doing particular, convenient things. Always feel free to use your ingenuity to avoid an algorithm that is not working for you, or seems to you to be unduly complex.

A word to the wise

We've done a lot of field-testing of the "hands-on" material in this book. Our instructions seem to be, on the whole, quite comprehensible to most readers. However, there are two basic types of error that people seem prone to make in carrying out our instructions.

Material error

In doing mathematics, it is absurd to specify the quality of paper on which the mathematics should be done. However, when we describe to you how to make mathematical models, we must insist that the choice of material is not arbitrary. Instructions for making models that are easily constructed using gummed mailing tape are unlikely to be effective if a strip of paper taken from an exercise book is used instead. Sometimes it may be merely a question of the finished model not being sufficiently sturdy, but it may even be true that the instructions simply

17

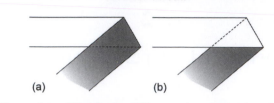

Figure 2.1   What is the difference between (a) and (b)?

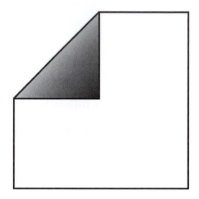

Figure 2.2   A square with a corner folded down.

cannot be carried out with inappropriate materials. Exercise your own initiative in choosing which models to make but not in your choice of material (except within very narrow limits).

Geometrical error

Look carefully at the two illustrations in Figure 2.1. Do you see a difference? If not, *look again!*

Notice that in (a) the portion of the strip going in the downward direction is *on top* of the horizontal part of the strip; whereas in (b) that portion is *underneath* the horizontal part of the strip. You will save yourself a great deal of time and effort if you will accustom yourself to looking very carefully at the illustrations, especially with respect to this distinction.

When a strip of paper is folded along a crease line, we indicate the revealed part of the *back* of the paper by shading as shown in Figure 2.2.

As we implied when constructing the flexagons in Chapter 1, there are at least 2 ways to turn a piece of paper over. We used a symbol whose orientation tells us which way something is to be turned over, as shown in Figure 2.3. Assume the square is a transparent square of plastic. Note that in (a) the orientation of the symbol tells you to flip the square over a horizontal axis along the bottom of the

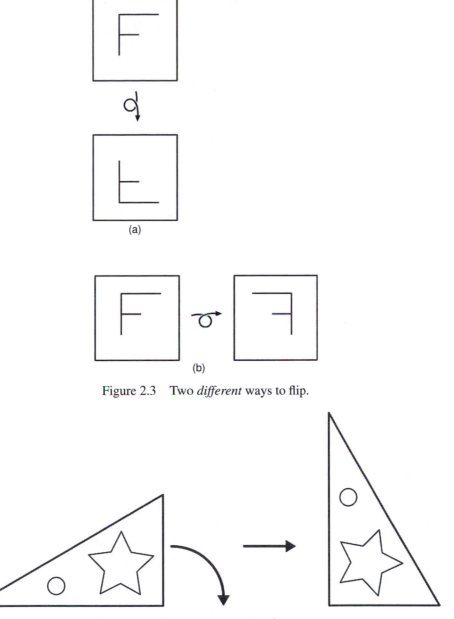

Figure 2.3 Two *different* ways to flip.

Figure 2.4 Showing the result of a move.

square; whereas in (b) the orientation of the symbol tells you to flip the square over a vertical axis along the right-hand side of the square.

In Figure 2.4 the heavy right-pointing arrow indicates that by performing the move on the left-hand figure (rotating the entire figure 90° in a clockwise direction about the right angle), we obtain the right-hand figure.

2.2 Some ancient threads

As is well-known, roundabout 350 B.C. the Greeks were fascinated with the idea of constructing regular N-gons with Euclidean tools (straightedge and compass). They were successful in constructing regular N-gons for $N = 2^c N_0$, with $c \geq 1$ and $N_0 = 1, 3, 5$, or 15, and $N \geq 3$ (so that the polygon will exist). Naturally the Greeks would have liked to answer the question for every N, but, in fact, it seems that no further progress was made until about 2000 years later, when Gauss (1777–1855) completely settled the question by proving that a Euclidean construction of a regular N-gon is possible *if and only if* N is of the form

$$N = 2^c \times \text{(a product of } distinct \text{ Fermat primes).}$$

A Fermat number has the form $F_n = 2^{2^n} + 1$. If it is prime it is called a **Fermat prime**.

Gauss's remarkable theorem tells us when a Euclidean construction is possible, provided we know which Fermat numbers are prime – which we don't! However, we do know that the following Fermat numbers are prime:

$$F_0 = 3, \quad F_1 = 5, \quad F_2 = 17, \quad F_3 = 257, \quad \text{and} \quad F_4 = 65537.$$

The great Swiss mathematician Euler (1707–1783) showed that $F_5 (= 2^{32} + 1)$ is not prime and, although many composite Fermat numbers have subsequently been identified, to this day no other Fermat numbers have proven to be prime, beyond those listed above. Thus, even with Gauss's contribution, a Euclidean construction of a regular N-gon is known to exist for very few values of N; and, even for these N, we do not know, in all cases, an explicit construction.

In the middle of the twentieth century one of the authors (JP) discovered a systematic folding procedure (see [55]), using straight strips of paper (like adding machine tape), that produced some regular polygons to any desired degree of accuracy. In fact, JP's paper-folding result came about by trying to construct the hexaflexagons of Chapter 1. Shortly thereafter she began working with PH and eventually they described a systematic method of producing a regular b-gon, to any desired degree of accuracy, for *any* integer $b \geq 3$. We note here that our objective is distinctly different from what the Greeks and Gauss were attempting to do, in that we systematically fold straight strips of paper, as will be described briefly below, and, once sufficient convergence has taken place, we use the folded strip to construct an *approximation* to the desired regular polygon, or star polygon, which we will define later.

Thus we redefined the Greeks' original problem. We agreed that we would be content to produce *approximations* to regular b-gons as long as we could depend on the error constantly becoming smaller. This seems resonable since Euclidean constructions are only perfect in the mind – after all, what is actually produced is a function of how sharp your pencil is, how steady you hold the compass, and how carefully you place the straightedge. Thus, even with Euclidean constructions, there are inevitable inaccuracies, due to human error. As you will soon see, in the systematic folding procedures which we use, every correct fold that is made always cuts any previous error in half – and, as you would expect, this produces very respectable (even if not completely perfect!) regular polygons.

2.3 Folding triangles and hexagons

Here is a sequence of illustrations that will enable you to fold a strip of equilateral triangles. To get the most out of this be sure to start with an angle x_0 that is *not* $\frac{\pi}{3}$ (that is, not 60°) as shown in frame 3 of Figure 2.5.

Continue folding to make a string of triangles as long as you need. Notice two things. First, the folding process goes UP, DOWN, UP, DOWN..., and we abbreviate it to $UDUDUD\ldots$ or U^1D^1, and sometimes refer to this folded strip as U^1D^1-*tape*. Second, although the first few triangles may be a bit irregular, the triangles formed always become more and more regular; that is, the angle between the last fold line and the edge of the tape gets closer and closer to $\frac{\pi}{3}$. When you use these triangles for constructing models, it is very safe to throw away the first 10 triangles and then to assume the rest of the triangles will be close enough to use for constructing anything that requires equilateral triangles.

Why do we get equilateral triangles?

To see what is happening all you need to do is to assume the first angle $x_0 = \frac{\pi}{3} + \epsilon$ (here, at least, ϵ is *not* restricted to positive values, it may be positive, zero, or *negative*!). Then following the first fold line to the top of the tape we see that the next angle, x_1, formed by a downward fold line, shown in frame 5 of Figure 2.5, would have to satisfy the equation $2x_1 + \frac{\pi}{3} + \epsilon = \pi$, from which it follows that $x_1 = \frac{\pi}{3} - \frac{\epsilon}{2}$. In the same way we can see that the new angle, x_2, created by the upward fold line, shown in frame 7, would have to satisfy the equation $2x_2 + x_1 = 2x_2 + \frac{\pi}{3} - \frac{\epsilon}{2} = \pi$, from which it follows that $x_2 = \frac{\pi}{3} + \frac{\epsilon}{2^2}$. Continuing in this way we find that $x_k = \frac{\pi}{3} + (-1)^k \frac{\epsilon}{2^k}$. This tells us that every time we make a correct fold the error is cut in half and changes sign. So after 10 correct fold lines the error must be smaller than $\frac{\epsilon}{1024}$!

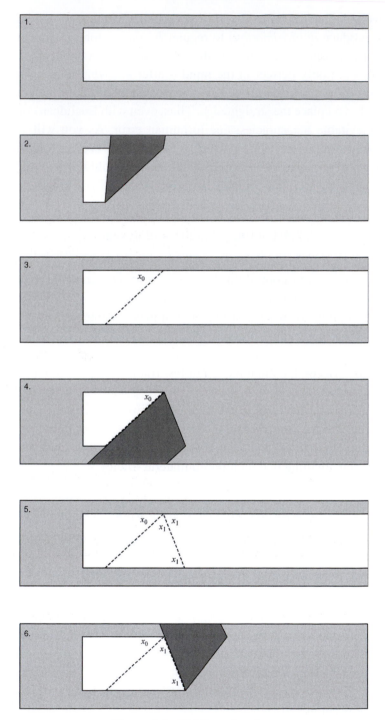

Figure 2.5   Folding triangles.

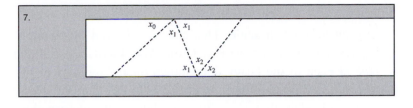

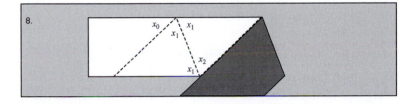

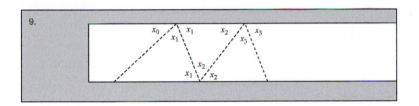

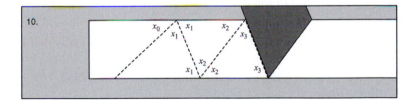

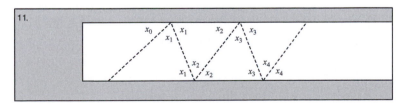

Figure 2.5   (*cont.*)

Constructing bigger triangles

Suppose you want a *bigger* triangle. This can be achieved by taking a strip of about 30 triangles and executing the FAT (**F**old-**A**nd-**T**wist) algorithm along the top of the tape at each of the heavy dots shown on the top part of Figure 2.6. The FAT algorithm is illustrated very precisely in Figure 2.6. We advise you to master this systematic algorithm, since it is used later in the construction of other regular polygons. The explanation of *why* the FAT algorithm works is given in Section 3.2.

Constructing hexagons

We can, of course, construct a hexagon, as we did in Chapter 1, by using the fact that a regular hexagon may be decomposed into 6 equilateral triangles. But there are other ways to construct the hexagon from the strip of equilateral triangles that are useful to know later on (when we want to construct regular N-gons of the form $2^n N_0$, where N_0 is any odd number). First, begin with a strip of folded equilateral triangles and make the *secondary* fold lines as shown in frames 2 and 3 of Figure 2.7 by bisecting the angle between the downward fold line and the top edge. Do this at every top vertex indicated by the arrow as shown in frame 4.

Now perform the FAT algorithm at each vertex along the top of the tape indicated by the arrows in frame 4 to obtain a regular 6-gon (with a hole in the center), as shown in Figure 2.8.

An even bigger hexagon can be obtained by increasing the distance between the successive vertices along the top of the tape at which you make your secondary folds. The hexagon is then formed, as before, by performing the FAT algorithm at 6 successive vertices equally spaced along the top of the tape. An example is shown in Figure 2.9, where secondary lines are folded at every other vertex along the top of the tape, and then the FAT algorithm is executed at those vertices using the secondary fold lines.

By now you can surely guess how you might construct 12-gons, 24-gons, etc. In fact, in theory you should be able to construct any $2^n 3$-gon by inserting enough secondary fold lines. The difficulty will come, of course, from trying to bisect an angle so many times. However, we would like to point out that this construction would also be a problem with a straightedge and compass.

2.4 Does this idea generalize?

Having had so much success with folding triangles it is very natural to ask what would happen if you folded twice at each vertex. To answer this question you

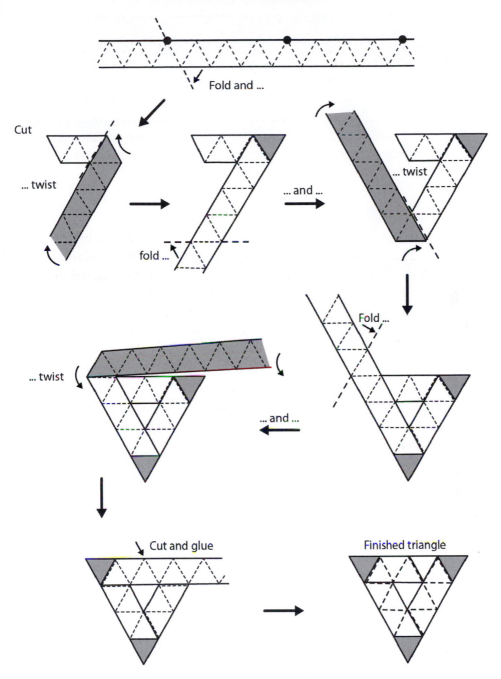

Figure 2.6   Constructing a FAT triangle.

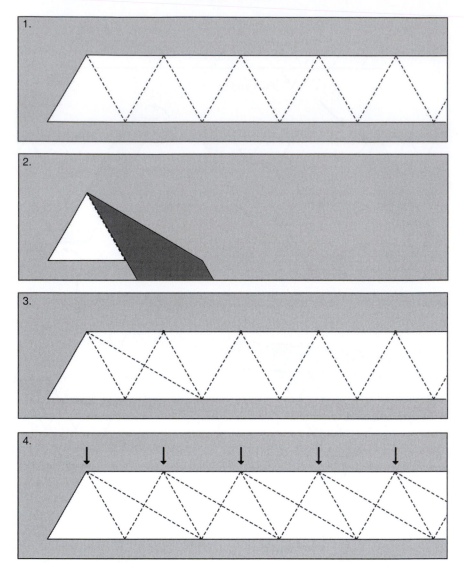

Figure 2.7   Preparing the $U^1 D^1$-tape for constructing a hexagon.

might experiment (as we did in the beginning) by following the folding procedure in Figure 2.10.

Notice that the tape, which we call $U^2 D^2$-tape (or, equivalently, $D^2 U^2$-tape) seems to be getting more and more regular – the successive long lines are becoming closer and closer to each other in length, and so are the successive short lines. The smallest angle on the tape seems to be approaching some fixed value. But what is it? To answer that question we suggest you take a piece of this tape, throw away

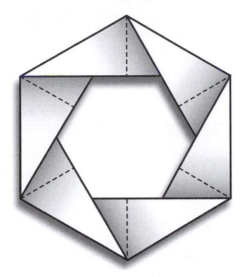

Figure 2.8　A FAT 6-gon.

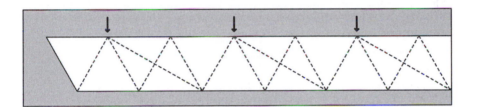

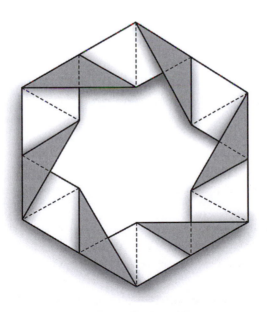

Figure 2.9　A bigger FAT 6-gon.

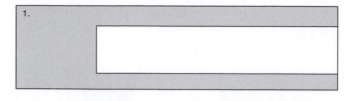

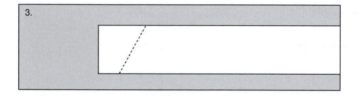

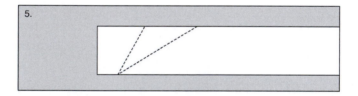

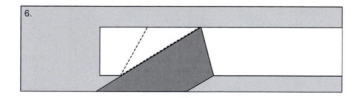

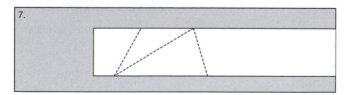

Figure 2.10   Folding UP UP DOWN DOWN . . . or U^2D^2.

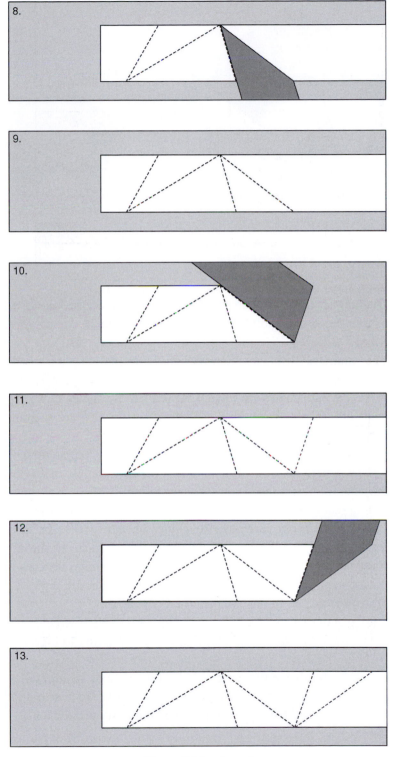

Figure 2.10   (*cont.*)

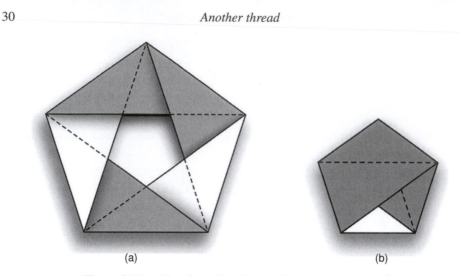

(a)   (b)

Figure 2.11   (a) A long-line 5-gon. (b) A short-line 5-gon.

the first 10 triangles, and play with it. Try folding it on successive long lines. Then try folding it on successive short lines. See if you can make the two polygons shown in Figure 2.11.

From the geometry of the situation we can figure out what the smallest angle on this U^2D^2-tape is approaching. Notice that in the top isosceles triangle of Figure 2.11(a) the base angles are equal. Let us call these angles α. Then, since the interior angle of a regular 5-gon is $\frac{3\pi}{5}$ we know that $2\alpha + \frac{3\pi}{5} = \pi$, from which it follows that $\alpha = \frac{\pi}{5}$.

There's more! If the FAT algorithm is executed at every vertex along the top of the folded tape, where the downward fold line makes an angle of $\frac{\pi}{5}$ with the top edge of the tape, we obtain the 5-gon shown in Figure 2.12.

By inserting secondary fold lines, just as we did with the U^1D^1-tape to produce FAT 6-gons, we can insert secondary fold lines on the U^2D^2-tape to enable us to produce FAT 10-gons. Figure 2.13 shows, in (a), a piece of the U^2D^2-tape on which secondary fold lines have been made at successive vertices along the top of the tape making an angle of $\frac{\pi}{5}$ to produce two adjacent angles of size $\frac{\pi}{10}$. Figure 2.13(b) shows part of the completed FAT 10-gon.

By inserting more secondary fold lines one can, in theory, construct regular 2^n5-gons, for any n. We would only be limited by our ability to repeatedly bisect angles.

As a reward for following these rather long constructions we will tell you about a particularly easy way to obtain a single pentagon. Just take a strip of paper and begin to tie an over-hand knot. As you pull the knot tight, press it flat as shown in Figure 2.14.

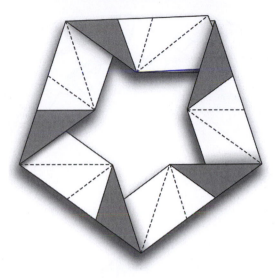

Figure 2.12   A FAT 5-gon.

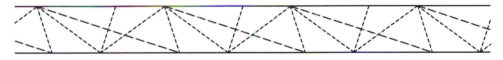

(a)

(b)

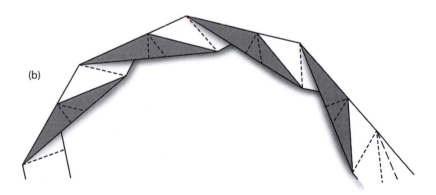

Figure 2.13   Constructing a FAT 10-gon.

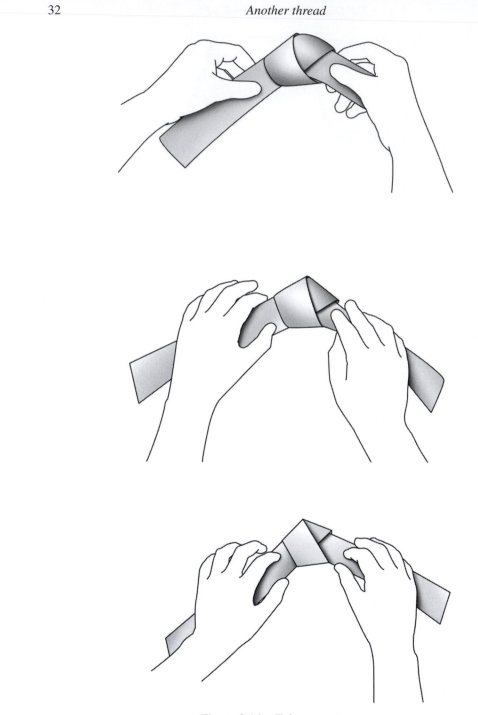

Figure 2.14   Tying a pentagon.

Table 2.1 *Loooking for a general pattern.*

By folding tape	and executing the FAT algorithm at equally spaced intervals along the top edge of the folded tape, we obtain a regular polygon having
$U^1 D^1$	3 sides
$U^2 D^2$	5 sides
$U^3 D^3$	? sides (make a guess)
$\vdots$	$\vdots$
$U^n D^n$	? sides (make a guess)

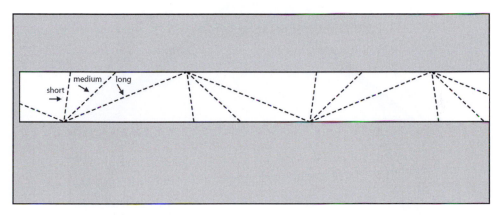

Figure 2.15   The beginning of a $U^3 D^3$-tape.

Is there a general pattern to all this?

So far in this chapter we have discussed a systematic folding procedure, where we make the same number of folds at the top of the tape as at the bottom of the tape. Furthermore, each of the fold lines bisects the angle, on the right, between the last fold line and an edge of the tape. Thus, as you may observe, all new fold lines will "go from left to right," sloping *up* if they are produced by an UP fold and sloping *down* if they are produced by a DOWN fold. If you keep this observation in mind, you can then simply "read off" the folding instructions from any folded strip of tape.

Let us review our results by placing them in a table. See if you can guess a general rule in Table 2.1.

Let us give you just one bit of information. The correct answer to the first question above is *not* 7 (but that is the most popular *wrong* answer!).

We're sure that by now you don't need a step-by-step diagram to know how to fold the $U^3 D^3$-tape. So we just show you in Figure 2.15 how the beginning folds of this tape would look.

Table 2.2 *What we know.*

When the number of bisections is	We can construct regular polygons with
1	3 sides
2	5 sides
3	9 sides

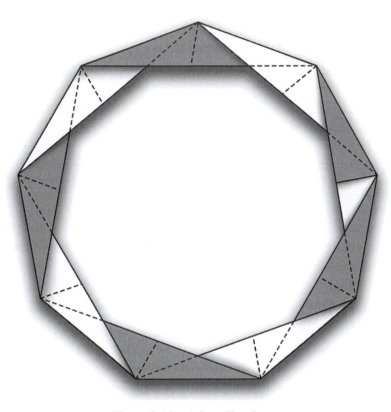

Figure 2.16   A *long*-line 9-gon.

You might like to experiment with a piece of $U^3 D^3$-tape (remembering to throw away the first few triangles). Fold it on successive short lines, or on successive medium lines, or on successive long lines. Each time you should get a polygon with the same number of sides. Or, you may execute the FAT algorithm on the long lines. Figures 2.16–2.19 show the resulting polygons that may be obtained from the $U^3 D^3$-tape.

Now we know that the answer to the first question in Table 2.1 is 9. So, to put what we know in tabular form, we construct Table 2.2.

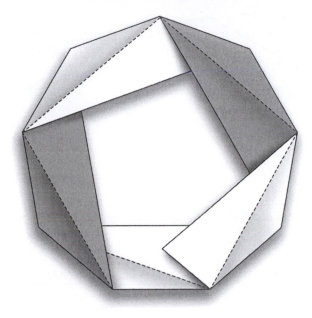

Figure 2.17   *A medium*-line 9-gon.

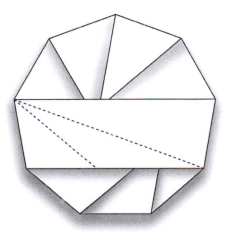

Figure 2.18   *A short*-line 9-gon.

Now, since we are heavily involved with *bisecting* angles we might suspect these numbers have something to do with the powers of 2. Ahah! The numbers 2, 4, and 8 are closely related to the sequence 3, 5, and 9. In fact, we see that

$$3 = 2^1 + 1$$
$$5 = 2^2 + 1$$
$$9 = 2^3 + 1.$$

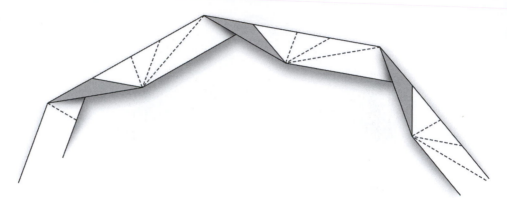

Figure 2.19   Part of a FAT 9-gon, constructed by performing the FAT algorithm on *long* lines of the $U^3 D^3$-tape.

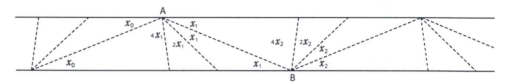

Figure 2.20   The beginning part of the $U^3 D^3$-tape.

So now we should suspect that if we fold $U^n D^n$, we can use that tape to construct regular $(2^n + 1)$-gons. To see that this is true we can give an *error-correction* proof analogous to that which we gave in the case $n = 1$. We demonstrate this, using Figure 2.20, in the particular but not special case when $n = 3$.

Assuming the first fold line in Figure 2.20 creates an angle of $x_0 = \frac{\pi}{9} + \epsilon$ (where, again, ϵ is completely free to be any real number, and we have inside information that ϵ particularly enjoys being negative!), then, at point A at the top of the tape we have $\frac{\pi}{9} + \epsilon + 2^3 x_1 = \pi$, from which it follows that $x_1 = \frac{\pi}{9} - \frac{\epsilon}{2^3}$. In a similar way we go to point B at the bottom of the tape to see that $\frac{\pi}{9} - \frac{\epsilon}{2^3} + 2^3 x_2 = \pi$, from which it follows that $x_2 = \frac{\pi}{9} + \frac{\epsilon}{(2^3)^2}$. Following this line of reasoning we see that, at the kth stage, we have

$$x_k = \frac{\pi}{9} + (-1)^k \frac{\epsilon}{(2^3)^k}.$$

Note that this verifies that every time a correct fold is made the error is cut in half (count the number of fold lines before the pattern repeats to see why you have the factor of 2^3 appearing in the denominator of the error term), and every time you complete the folding on one edge of the tape the error on the other edge of the tape will change sign.

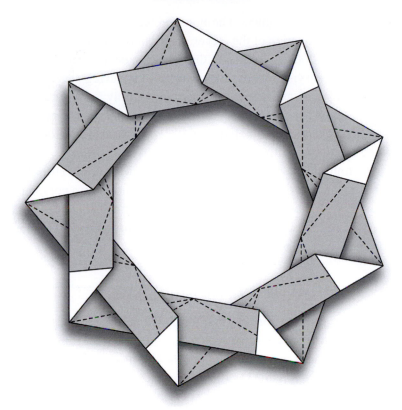

Figure 2.21   A regular $\{\frac{9}{2}\}$-gon, formed by performing the FAT algorithm on *medium* lines of the U^3D^3-tape.

2.5 Some bonuses

As we have discovered in other contexts, mathematics is generous, often giving us much more than we originally asked for.

Notice that numbers of the form $2^n + 1$ include $2^4 + 1 = 17$. Thus we now know how we may construct Gauss's beloved 17-gon. Simply fold a strip of tape using the U^4D^4-procedure and then apply the FAT algorithm (on the longest line) at 17 equally spaced vertices along the top of the tape. Alternatively, fold consistently on one of the 4 different kinds of crease lines, while leaving the other crease lines flat.

In theory, we could use this method to construct the 33-gon (since $2^5 + 1 = 33$), the 65-gon (since $2^6 + 1 = 65$), the 129-gon (since $2^7 + 1 = 129$), the 257-gon, and so on. In principle, we could even construct the 65 537-gon (by folding $U^{16}D^{16}$). However, as we have said, it is very difficult in practice to fold either up or down in the prescribed manner more than 4 times.

Actually, we can do even more with the U^3D^3-tape. You may have already noticed in your own investigations that it is possible to perform the FAT algorithm on the U^3D^3-tape along the medium-length lines. Figure 2.21 shows the *star* polygon that is obtained by this procedure. The notation we use for this polygon (that is, the $\{\frac{9}{2}\}$-gon), is an adaptation of the clever notation used by the great geometer H. S. M. Coxeter (1907–2003) – see [8]. In this case, the denominator 2 indicates that the top edge of the tape visits every *second* vertex of some regular convex polygon, and the numerator 9 indicates that the regular convex polygon we are talking about has *nine* vertices (and hence 9 sides). All our previous polygons could be denoted, if we wished, by this fractional notation, with 1 in the denominator, but as usual we follow the custom of writing only the numerator in this case.

The strict definition of a polygon (or polygonal path) does not allow the sides of the polygon to cross each other. Thus a star polygon is not truly a polygon – but that doesn't mean it is not a beautiful geometrical figure of considerable mathematical interest. We will call these polygons *quasi-regular polygons*, to distinguish them from the ordinary convex polygons.

Uses for our folded tape

Since we now have a satisfactory method for constructing good approximations of regular 3- and 5-gons and we already know how to construct squares, we can use our strips of folded paper to make many polyhedral models. In Chapter 5 we give instructions for constructing the regular convex polyhedra known as the Platonic solids. In Chapter 6 we give instructions for using strips of triangles to construct rotating rings of tetrahedra and certain regular deltahedra with just one strip of triangles. In Chapter 9 we describe how to braid the Platonic solids from straight strips of the folded tape. In Chapter 12, on the other hand, we give instructions for using the strips of triangles to construct what we call *collapsoids*, which are non-convex models which fold up in accordion style. But, first, we want to continue the paper-folding thread a little further, in Chapter 3. There we will also show you *why* we had to invent the FAT algorithm (notice that it wasn't necessary to use the FAT algorithm to get any of the polygons we have so far discussed).

3

More paper-folding threads – 2-period paper-folding

3.1 Some basic ideas about polygons

Since we are shortly going to take on the task of constructing less familiar polygons, we need to get a little bit more technical. So we discuss next a key result in the Euclidean geometry of the plane which is absolutely basic to what follows concerning the construction of polygons.

Given any polygon (it need not even be convex, let alone regular), we claim that the sum of its exterior angles is 2π radians (just assume π radians $= 180°$ if you don't know about radians). A nice way of seeing this is, oddly enough, to pretend to be extremely short-sighted. The diagram shown in Figure 3.1(a) looks to anyone with serious myopia like that shown in Figure 3.1(b). Or, alternatively, imagine you walk so far away from the figure in Figure 3.1(a) that you can barely still see it. Then it will appear as shown in Figure 3.1(b).

The following facts are immediate consequences:

(1) The sum of the interior angles of an N-gon is $(N - 2)\pi$.
(2) The exterior angles of a *regular* N-gon are each $\frac{2\pi}{N}$.
(3) The interior angles of a *regular* N-gon are each $\left(\frac{N-2}{N}\right)\pi$.

3.2 Why does the FAT algorithm work?

We used, without explaining why it worked, the FAT algorithm in Chapter 2. For the polygons we have constructed so far we didn't actually *need* to use the FAT algorithm to obtain the polygon; this was because the geometry of the $U^n D^n$-tape allowed us to obtain the regular $(2^n + 1)$-gon if it was folded on successive lines of any fixed length (and there were always n such lengths). So, in those cases, the FAT algorithm just gave us a bonus $(2^n + 1)$-gon.

However, we aren't always going to be so lucky – and that is why the FAT algorithm needed to be invented. Here is the explanation of *why* it works in a very general setting that produces regular star $\left\{\frac{b}{a}\right\}$-gons.

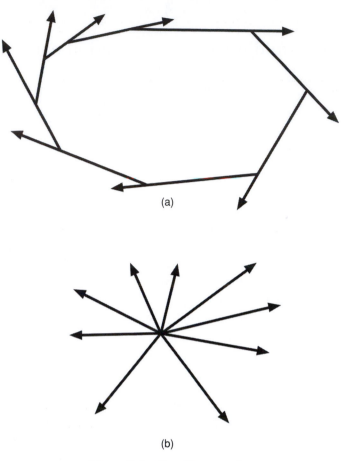

(a)

(b)

Figure 3.1   Proof by myopia.

Assume that we have a straight strip of paper that has certain vertices marked on its top and bottom edges and which also has *creases* or *folds* along straight lines emanating from vertices at the top edge of the strip. Further assume that the creases at those vertices labeled A_{nk}, $n = 0, 1, 2, \ldots$ (see Figure 3.2), which are on the top edge, form identical angles of $\frac{a\pi}{b}$ with the top edge, with an identical angle of $\frac{a\pi}{b}$ between the crease along the lines $A_{nk}A_{nk+2}$ and the crease along $A_{nk}A_{nk+1}$ (as shown in Figure 3.2(a)). Suppose further that, for each k, the vertices A_{nk}, $n = 0, 1, 2, \ldots$, are equally spaced. If we fold this strip on $A_{nk}A_{nk+2}$, as shown in Figure 3.2(b), and then twist the tape so that it folds on $A_{nk}A_{nk+1}$, as shown in Figure 3.2(c), the direction of the *top edge* of the tape will be rotated through an angle of $2\left(\frac{a\pi}{b}\right)$. As you know, from Chapter 2, we call this process of <u>F</u>olding <u>A</u>nd <u>T</u>wisting the *FAT algorithm* (see any of [26–30, 33, 35, 38]).

Now consider the equally spaced vertices A_{nk} along the top of the tape, with k fixed and n varying. If the FAT algorithm is performed on a sequence of angles,

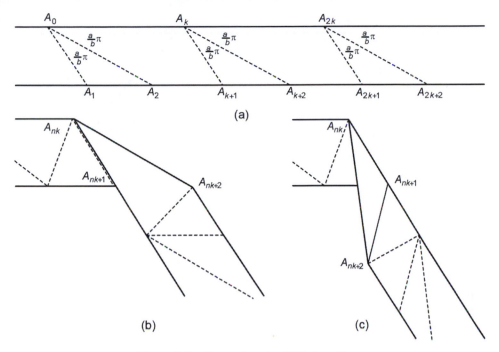

Figure 3.2 Executing the FAT algorithm.

each of measure $\frac{a\pi}{b}$, at the vertices given by $n = 0, 1, 2, \ldots, b-1$, then the top of the tape will have turned through an angle of $2a\pi$. Thus the vertex A_{bk} will come into coincidence with A_0; and the top edge of the tape will have visited every ath vertex of a bounding regular convex b-gon, thus determining a regular star $\left\{\frac{b}{a}\right\}$-gon. As an example, recall the $\left\{\frac{9}{2}\right\}$-gon in Figure 2.21, $a = 2$ and $b = 9$.

As was touched on in Chapter 2, we are making a slight adaptation of the Coxeter notation for star polygons (see [8]), so that when we refer to a regular star $\left\{\frac{b}{a}\right\}$-gon we mean a connected sequence of edges that visits every ath vertex of a regular convex b-gon. Thus our N-gon is the special star $\left\{\frac{N}{1}\right\}$-gon. When viewing a convex polygon this way we may well use a lowercase letter instead of N.

Figure 3.3 illustrates how a suitably creased strip of paper may be folded by the FAT algorithm to produce a regular p-gon. In Figure 3.3 we have written V_k instead of A_{nk}, since it is more natural in this particular context.

Let us now illustrate how the FAT algorithm may be used to fold a regular convex 8-gon. Figure 3.4(a) shows a straight strip of paper on which the dotted lines indicate certain special exact crease lines. In fact, these crease lines occur at equally spaced intervals along the top of the tape so that the angles occurring at the top of each vertical line are (from left to right) $\frac{\pi}{2}$, $\frac{\pi}{4}$, $\frac{\pi}{8}$, and $\frac{\pi}{8}$. Figuring out how to fold a strip of tape to obtain this arrangement of crease lines is unlikely to cause the

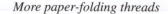

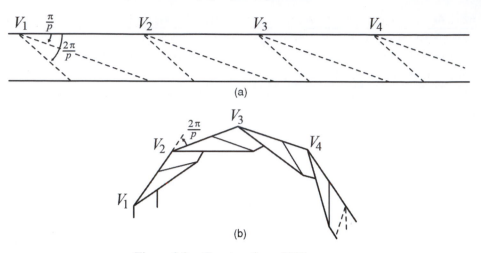

(a)

(b)

Figure 3.3   Constructing a FAT p-gon.

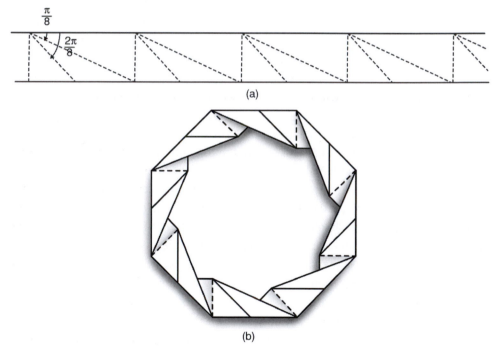

(a)

(b)

Figure 3.4   Constructing the FAT 8-gon.

reader any difficulty, but complete instructions are given in [33]. Our immediate interest is focused on the observation that this tape has, at equally spaced intervals along the top edge, adjacent angles each measuring $\frac{\pi}{8}$, and we can therefore execute the FAT algorithm at 8 consecutive vertices along the top of the tape to produce an exact regular convex 8-gon shown as Figure 3.4(b). (Of course, in constructing the model one would cut the tape on the first vertical line and glue a section at the end to the beginning so that the model would form a closed polygon.)

Notice that the tape shown in Figure 3.4(a) also has suitable crease lines that make it possible to use the FAT algorithm to fold a regular convex 4-gon. We leave this as an exercise for the reader and turn to a more challenging construction, the regular convex 7-gon.

3.3 Constructing a 7-gon

Now, since the 7-gon is the first regular polygon that we encounter for which we do not have available a Euclidean construction (nor does anybody else), we are faced with a real difficulty in creating a crease line making an angle of $\frac{\pi}{7}$ with the top edge of the tape. Furthermore, since 7 isn't of the form $2^n + 1$, we can't obtain suitable tape from the $U^1 D^1$-folding procedure. We're in real trouble here, so we've got to think of another approach.

We proceed by adopting a general policy which we call our ***optimistic strategy***. Assume that we *can* crease an angle of $\frac{2\pi}{7}$ (certainly we can come close) as shown in Figure 3.5(a). Given that we have the angle of $\frac{2\pi}{7}$, it is then a trivial matter to fold the top edge of the strip DOWN to bisect this angle, producing two adjacent angles of $\frac{\pi}{7}$ at the top edge as shown in Figure 3.5(b). (We say that $\frac{\pi}{7}$ is the ***putative*** angle on this tape.) Then, since we are content with this arrangement, we go to the bottom of the tape where we observe that the angle to the right of the last crease line is $\frac{6\pi}{7}$ – and we decide, as paper-folders, that we will always avoid leaving even multiples of π in the numerator of any angle next to the edge of the tape, so we bisect this angle of $\frac{6\pi}{7}$, by bringing the bottom edge of the tape UP at an angle of $\frac{3\pi}{7}$ to coincide with the last crease line and creating the new crease line sloping up shown in Figure 3.5(c). We settle for this (because we are content with an odd multiple of π in the numerator) and go to the top of the tape where we observe that the angle to the right of the last crease line is $\frac{4\pi}{7}$ – and, since we have decided against leaving an even multiple of π in any angle next to an edge of the tape, we are forced to bisect this angle twice, each time bringing the top edge of the tape DOWN to coincide with the last crease line, obtaining the arrangement of crease lines shown in Figure 3.5(d). But now we notice something miraculous has occurred! If we had really started with an angle of exactly $\frac{2\pi}{7}$, and if we now continue introducing crease lines by repeatedly folding the tape DOWN TWICE at the top and UP ONCE at the bottom, we get precisely what we want; namely, pairs of adjacent angles, measuring $\frac{\pi}{7}$, at equally spaced intervals along the top edge of the tape. Let us call this folding procedure the ***$D^2 U^1$-folding procedure*** (or, more simply – and especially when we are concerned merely with the related

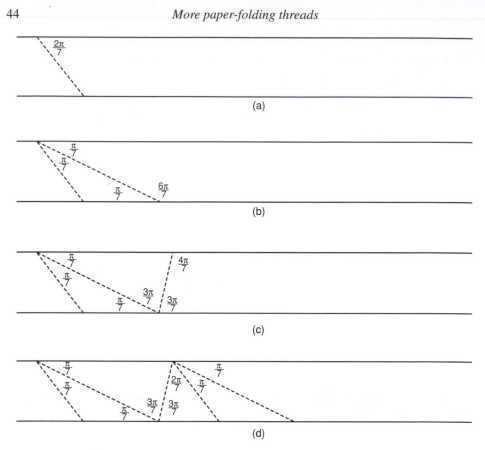

Figure 3.5   Devising the folding scheme to produce FAT 7-gons.

number theory – the (**2**, **1**)-*folding procedure*) and call the strip of creased paper it produces D^2U^1-*tape* (or, again more simply, (**2**, **1**)-*tape*). The crease lines on this tape are called the ***primary crease lines***.

The really interested paper-folder should, before reading further, get a piece of paper and fold an acute angle which you would call an approximation to $\frac{2\pi}{7}$. Then fold about 40 triangles using the D^2U^1-folding procedure as shown in Figures 3.5 and 3.6(a) as described above, throw away the first 10 triangles, and see if you can tell that the first angle you get between the top edge of the tape and the adjacent crease line is *not* $\frac{\pi}{7}$. Then try to construct the FAT 7-gon shown in Figure 3.6(b). You will then *believe* that the D^2U^1-folding procedure produces tape on which the smallest angle does, indeed, approach $\frac{\pi}{7}$, actually rather rapidly.

You might also try executing the FAT algorithm at every other vertex along the top of this tape to produce a regular $\left\{\frac{7}{2}\right\}$-gon. (Hint: Look at Figure 3.6(c).)

How do we *prove* that this evident convergence actually takes place? A direct approach is to admit that the first angle folded down from the top of the tape in

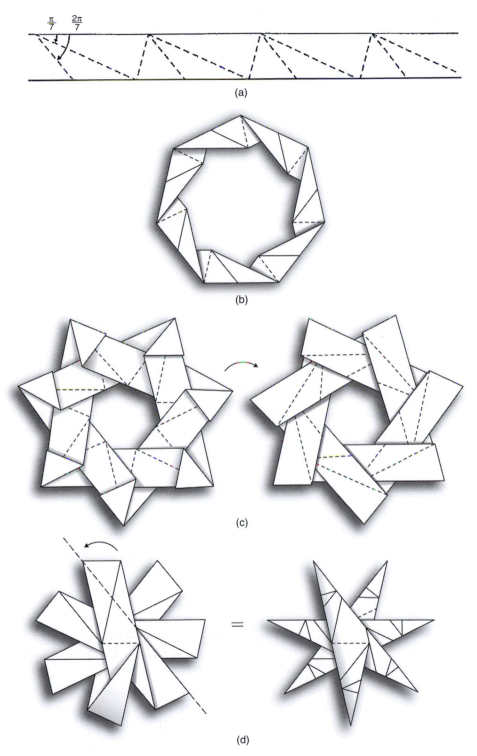

Figure 3.6   Three FAT 7-gons.

Figure 3.5(a) might not have been precisely $\frac{2\pi}{7}$. Then the bisection forming the next crease would make the two acute angles nearest the top edge in Figure 3.5(b) only approximately $\frac{\pi}{7}$; let us call them $\frac{\pi}{7} + \epsilon$ (where ϵ may be either positive or negative). Consequently the angle to the right of this crease, at the bottom of the tape, would measure $\frac{6\pi}{7} - \epsilon$. When this angle is bisected, by folding up, the resulting acute angles nearest the bottom of the tape, labeled $\frac{3\pi}{7}$ in Figure 3.5(c), would in fact measure $\frac{3\pi}{7} - \frac{\epsilon}{2}$, forcing the angle to the right of this crease line at the top of the tape to have measure $\frac{4\pi}{7} + \frac{\epsilon}{2}$. When this last angle is bisected twice by folding the tape down, the two acute angles nearest the top edge of the tape will each measure $\frac{\pi}{7} + \frac{\epsilon}{2^3}$. This makes it clear that, every time we repeat a $D^2 U^1$-folding on the tape, the error is reduced by a factor of 2^3.

We see that our ***optimistic strategy*** has paid off – by blandly *assuming* we have an angle of $\frac{2\pi}{7}$ at the top of the tape to begin with, and folding accordingly, we *get what we want* – successive angles at the top of the tape which, as we fold, rapidly get closer and closer to $\frac{\pi}{7}$, whatever angle we had, in fact, started with! We thus say that $\frac{\pi}{7}$ is the *putative* angle on this tape.

Figures 3.6(c), (d) show the regular $\left\{\frac{7}{2}\right\}$- and $\left\{\frac{7}{3}\right\}$-gons that are produced from the $D^2 U^1$-tape by executing the FAT algorithm on the crease lines that make angles of $\frac{2\pi}{7}$ and $\frac{3\pi}{7}$, respectively, with an edge of the tape (if the angle needed is at the bottom of the tape, as with $\frac{3\pi}{7}$, simply turn the tape over so that the required angle appears on the top). In Figures 3.6(c), (d) the FAT algorithm was executed on every other suitable vertex along the edge of the tape so that, in (c), the resulting figure, or its flipped version, could be woven together in a more symmetric way and, in (d), the excess could be folded neatly around the points.

It is now natural to ask:

(1) Can we use the same general approach used for folding a convex 7-gon to fold a convex N-gon with N odd, at least for certain specified values of N? If so, can we always prove that the actual angles on the tape really converge to the putative angle we originally sought?

(2) What happens if we consider general folding procedures perhaps with other periods, such as those represented by

$$D^3 U^3, \quad D^4 U^2, \quad \text{or} \quad D^3 U^1 D^1 U^3 D^1 U^1?$$

(The ***period*** is determined by the repeat of the *exponents*, so these examples have periods 1, 2, and 3, respectively.)

The answer to (1) is *yes* and we will show, in Chapter 7, an algorithm for determining a folding procedure that produces tape from which you can construct *any* regular $\left\{\frac{b}{a}\right\}$-gon, if a and b are odd with $a < \frac{b}{2}$. A partial answer to (2) appears in Section 7.2, but here we will simply note that an iterative folding procedure of this type will, in fact, always converge (what it converges *to* is quite another matter – but we will be telling you about that later in Section 7.1).

*3.4 Some general proofs of convergence

It turns out that, in answering the questions that arose in the last section, we need to make a straightforward use of the following theorem which is a special case of the well-known contraction mapping principle (see, e.g. [82]).

Theorem 3.1 *For any three real numbers a, b, and x_0, with a $\neq$ 0, let the sequence $\{x_k\}$, k = 0, 1, 2, . . . , be defined by the recurrence relation*

$$x_k + ax_{k+1} = b, \qquad k = 0, 1, 2, \ldots \tag{3.1}$$

Then, if $|a| > 1$, $x_k \to \frac{b}{1+a}$ as $k \to \infty$.

Proof. Set $x_k = \frac{b}{1+a} + y_k$. Then $y_k + ay_{k+1} = 0$. It follows that $y_k = \left(\frac{-1}{a}\right)^k y_0$. If $|a| > 1$, $\left(\frac{-1}{a}\right)^k \to 0$, so that $y_k \to 0$ as $k \to \infty$. Hence $x_k \to \frac{b}{1+a}$ as $k \to \infty$. Notice that y_k is the *error* at the kth stage, and that the absolute value of y_k is equal to $\frac{1}{|a|^k}|y_0|$. □

We point out that it is significant in Theorem 3.1 that neither the convergence nor the limit depends on the initial value x_0. This means, in terms of the folding, that the process will converge to the same limit no matter how we fold the tape to produce the first crease line – this is what justifies our *optimistic strategy*! And, as we have seen in our examples, and, as we will soon demonstrate in general, the result of the theorem tells us that the convergence of our folding procedure is rapid, since, in all cases, $|a|$ will be a positive power of 2.

Let us now look again at the general 1-period folding procedure $D^n U^n$. A typical portion of the tape would appear as illustrated in Figure 3.7(a). Then, if the folding process had been started with an arbitrary angle u_0, we would have, from the kth stage,

$$u_k + 2^n u_{k+1} = \pi, \qquad k = 0, 1, 2, \ldots \tag{3.2}$$

Equation (3.2) is of the form (3.1), so it follows from the argument of Theorem 3.1 that

$$u_k \to \frac{\pi}{2^n + 1} \qquad \text{as} \quad n \to \infty \tag{3.3}$$

Furthermore, we now see from (3.3) that, if the original fold differed from the *putative* angle of $\frac{\pi}{2^n+1}$ by an error of ϵ_0, then the error at the kth stage of the

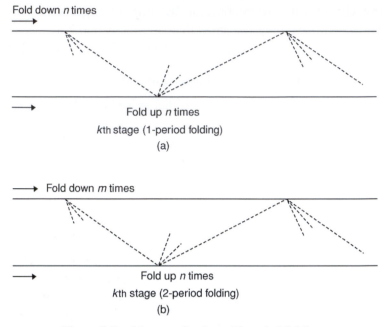

Fold down *n* times

Fold up *n* times

*k*th stage (1-period folding)

(a)

Fold down *m* times

Fold up *n* times

*k*th stage (2-period folding)

(b)

Figure 3.7   *k*th stage for 1- and 2-period folding.

$D^n U^n$-folding procedure would be given by

$$|\epsilon_k| = \frac{|\epsilon_0|}{2^{nk}}.$$

A particularly interesting case occurs when $n = 3$; for then we approximate the regular 9-gon, whose non-constructibility by Euclidean tools is very closely related to the non-trisectibility of an arbitrary angle.

But, as we have noted, the case $N = 2^n + 1$ is atypical, since we may construct $(2^n + 1)$-gons from our folded tape by special methods (not involving the FAT algorithm), in which, however, the top edge does not describe the polygon, as it does in the FAT algorithm. Figure 3.8 shows how the $D^2 U^2$-tape shown in part (a) may be folded along just the short lines of the creased tape to form the *outline* of a regular pentagon shown in (b), and along just the long lines of the creased tape to form the *outline* of the slightly larger regular pentagon shown in (c); and, finally, we show in (d) the regular pentagon formed by an edge of the tape when the FAT algorithm is executed on the long lines of the $D^2 U^3$-tape.

Recall from Section 2.3 that, if we already know how to construct regular N-gons, then, by the insertion of secondary crease lines, we may construct $2^c N$-gons. If, for example, we wished to construct a regular 10-gon then we take the $D^2 U^2$-tape (which you may recall produced FAT 5-gons) and introduce a **secondary crease line** by bisecting each of the angles of $\frac{\pi}{5}$ next to the top (or bottom) edge

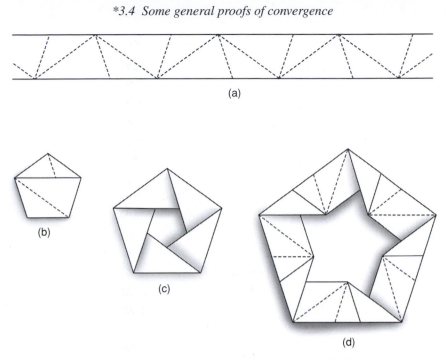

Figure 3.8   Three pentagons constructed from D^2U^2-tape.

of the tape. The FAT algorithm may be used on the resulting tape to produce the regular convex FAT 10-gon, as illustrated in Figure 3.9. It should now be clear how to construct a regular 20-gon, 40-gon, 80-gon, ...

Now we turn to the general **2-*period*** folding procedure, D^mU^n, which we may abbreviate as (m, n). (Recall that the tape that produced the regular 7-gon was a 2-period tape.) A typical portion of the 2-period tape, in the general case, is illustrated as shown in Figure 3.7(b). If the folding procedure had been started with an arbitrary angle u_0 at the top of the tape, and continues producing angles $u_1, u_2, \ldots$ at the top and $v_0, v_1, \ldots$ at the bottom, we would have, at the kth stage,

$$u_k + 2^n v_k = \pi,$$

$$v_k + 2^m u_{k+1} = \pi,$$

and hence it follows, by substituting $v_k = \pi - 2^m u_{k+1}$ into the top equation, that

$$u_k - 2^{m+n}u_{k+1} = \pi(1 - 2^n), \qquad k = 0, 1, 2, \ldots \qquad (3.4)$$

Thus, again using Theorem 3.1, we see that

$$u_k \to \frac{2^n - 1}{2^{m+n} - 1}\pi \qquad \text{as} \quad k \to \infty, \qquad (3.5)$$

$\frac{\pi}{10}\ \frac{2\pi}{10}$

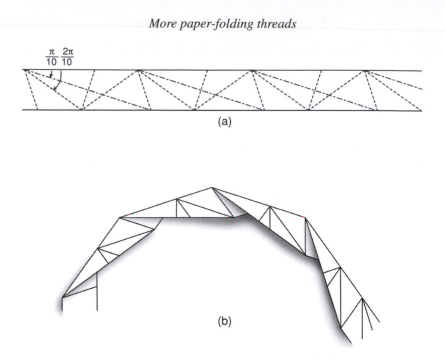

(a)

(b)

Figure 3.9   Constructing a 10-gon from D^2U^2-tape.

so that $\dfrac{2^n - 1}{2^{m+n} - 1}\pi$ is the *putative* angle at the top of the tape. Thus the FAT algorithm will produce, from this tape, a regular $\left\{\frac{b}{a}\right\}$-gon, where the fraction $\frac{b}{a}$ may turn out not to be reduced, with $b = 2^{m+n} - 1$, $a = 2^n - 1$. For example, when $n = 2$, $m = 4$, we would have $u_k \to \dfrac{2^2 - 1}{2^6 - 1}\pi = \dfrac{3}{63}\pi = \dfrac{\pi}{21}$.

By symmetry we infer that

$$v_k \to \frac{2^m - 1}{2^{m+n} - 1}\pi \qquad \text{as} \quad k \to \infty. \tag{3.6}$$

Furthermore, if we assume an initial error of ϵ_0, then Theorem 3.1 shows that the error at the kth stage (that is, when the (m, n)-folding has been done exactly k times) will be given by

$$\epsilon_k = \frac{\epsilon_0}{2^{(m+n)k}}. \tag{3.7}$$

Hence, we see that in the case of our D^2U^1-folding (Figures 3.5 and 3.6) any initial error ϵ_0 is, as we have already seen from our previous argument, reduced by a factor of 8 between consecutive stages. It should now be clear why we advised throwing away the first part of the tape – but, likewise, it should also be clear that it is never necessary to throw away very much of the tape. In practice, convergence is very rapid indeed, and if one made it a rule of thumb to always throw away the first 10 crease lines on the tape for any iterative folding procedure, it would turn out to be a very conservative rule.

We see that, however wonderful these results may be, they haven't, in fact, completely solved the problem of how to approximate *any* N-gon. For example, as you may verify, we would be unable to fold a regular 11-gon with either the 1- or 2-period folding procedure. So the question remains: ***How do we know which sequence of folds to make in order to produce a particular regular polygon with the FAT algorithm?*** We answer this question in Chapter 7, and turn now to some number theory connected with the 2-period folding procedure discussed in the next chapter.

4

A number-theory thread – Folding numbers, a number trick, and some tidbits

4.1 Folding numbers

This chapter contains some serious number theory. But, even if you aren't a number-theorist or do not have a special interest in number theory, we would encourage you to give this chapter some attention because it leads to surprising results that we could not have discovered had we not begun this exploration by folding paper! For example, the innocent-looking statement that $3 \times 7 = 21$ usually leads to two equivalent statements; namely that $21 \div 3 = 7$ and $21 \div 7 = 3$. But it turns out that, in an important and interesting sense, the first statement is very special while the second is merely one example of an infinite number of such statements.

Thus, even if you choose not to study the proofs of the theorems in this chapter in depth, you might get quite a bit of pleasure by seeing what the theorems say and looking at some examples of them.

Let us emphasize this point further. Many people – especially students – believe that a proof should be committed to memory, and try to do so. This is wrong. The purpose of a displayed proof is to convey the meaning of the statement of a theorem, and its significance. If it does not do that, it has failed to play its proper role in the student's education, however accurately it may be recorded in the student's memory. It should help the student to understand, and not just add to his, or her, difficulties.

Now we return to the mathematics. We saw in Section 3.4 that the $D^m U^n$-folding procedure produced a putative angle of $\dfrac{2^n - 1}{2^{m+n} - 1}\pi$ along the top of the tape that allowed us to produce (via the FAT algorithm) a regular s-gon, where

$$s = \frac{2^{m+n} - 1}{2^n - 1}, \qquad m \geq 1, \quad n \geq 1, \tag{4.1}$$

provided only that s is an integer; and that, in any case, it will produce a quasi-regular star $\left\{ \dfrac{2^{m+n} - 1}{2^n - 1} \right\}$-gon. We naturally now ask our first *number-theoretic*

52

question:

$$\text{\textbf{\textit{When is }} } \frac{2^{m+n} - 1}{2^n - 1} \text{ \textit{an integer?}}$$

   Let's be really ambitious! From the purely *number-theoretic* point of view there is no reason to confine attention to the number 2, in (4.1); this choice was dictated by our insistence on regarding the *bisection* of angles as a fundamental geometrical operation. Thus we will fix an arbitrary integer $t \geq 2$ as **base** and discuss rational numbers of the form $\frac{t^a - 1}{t^b - 1}$, for given positive integers a, b. Our first question thus becomes

$$\text{\textbf{\textit{When is }} } \frac{t^a - 1}{t^b - 1} \text{ \textit{an integer?}}$$

We will find, perhaps surprisingly, that the answer is independent of our choice of the base t, thus confirming the correctness of our strategy in making this generalization. In order to deal also with rational numbers of the form $\frac{t^a - 1}{t^b - 1}$, we prove the following theorem.

Theorem 4.1 *Let $d = \gcd(a, b)$. Then $\gcd(t^a - 1, t^b - 1) = t^d - 1$.*

Proof. Let § stand for $t - 1$ in base t. Then, writing numbers in base t, we see that[1]

$$\begin{aligned}
t^a - 1 &\stackrel{(t)}{=} \S\S \cdots \cdots \S \quad (a \text{ squiggles}), \\
t^b - 1 &\stackrel{(t)}{=} \S\S \cdots \S \quad\quad (b \text{ squiggles}).[2]
\end{aligned} \tag{4.2}$$

Now there is a beautiful algorithm for calculating gcd's, called the **Euclidean algorithm**. If we want to calculate $\gcd(a, b)$, where we assume $a > b$, we divide a by b, paying no attention to the quotient but recording the remainder as r_1; where $0 \leq r_1 < b$. If $r_1 \neq 0$, we divide b by r_1, recording the remainder as r_2; where $0 \leq r_2 < r_1$. If $r_2 \neq 0$, we divide r_1 by r_2, recording the remainder as r_3; and we continue in this way until we get, as eventually we must, a remainder $r_k = 0$. Then $\gcd(a, b) = r_{k-1}$. (If $k = 1$, interpret r_0 as b.)

   (The proof resumes after some examples of the Euclidean algorithm.)

---

[1] We write $\stackrel{(t)}{=}$ to indicate that the number on the right is written in base t.
[2] It may be helpful to recall the well-known case $t = 10$ where $\S = 9$. Thus, for example, $10^5 - 1 = 99999$.

As an example let us use the above description to calculate gcd(27 379, 341), and show you why the algorithm works. First, we execute the algorithm, writing out the successive steps as

$$\begin{aligned}
27\,379 &= 341\,Q_1 + 99 && (r_1 = 99) \\
341 &= 99\,Q_2 + 44 && (r_2 = 44) \\
99 &= 44\,Q_3 + 11 && (r_3 = 11) \\
44 &= 11\,Q_4 + 0 && (r_4 = 0).
\end{aligned}$$

So we claim that gcd(27 379, 341) $= r_3 = 11$. Second, to justify our claim, we observe, from the first equation, that *any* number which is a divisor of both 27 379 and 341 must also be a divisor of both 341 and 99; and conversely. Likewise, *any* number which is a divisor of both 341 and 99 must also be a divisor of both 99 and 44; and conversely. And so on. Now we can see that 11 is the greatest common divisor of 44 and 11. Hence the successive steps above show that

$$\gcd(27\,379, 341) = \gcd(341, 99) = \gcd(99, 44) = \gcd(44, 11) = 11.$$

See if you can figure out how to use your hand calculator to obtain the above sequence of equations about gcd's. Of course, as soon as you recognize the gcd, you can go straight to the final answer.

It is an interesting exercise to calculate the gcd of some consecutive Fibonacci, or Lucas, numbers.[3] You may notice that these calculations take a very large number of steps, and the answer is always somewhat the same. The general argument involves the observation that $\gcd(F_{n+2}, F_{n+1}) = \gcd(F_{n+1}, F_n)$, by the recurrence relation $F_{n+2} = F_{n+1} + F_n$, $n \geq 0$. Thus, working backwards, we see that, since $F_1 = 1$, it follows that the gcd of *any* two consecutive Fibonacci numbers must be 1. Of course, there is an almost identical result for Lucas numbers. The argument uses the recurrence relation $L_{n+2} = L_{n+1} + L_n$, $n \geq 0$, and the fact that $L_1 = 1$.

Continuing the proof of Theorem 4.1. It follows, from (4.2) in base t, remembering that r_1 is the remainder when we divide a by b, that, if we divide $t^a - 1$ by $t^b - 1$, we get a remainder of $t^{r_1} - 1$; and that the remainders are indeed successively

$$t^{r_2} - 1, \qquad t^{r_3} - 1, \ldots$$

Thus if the first zero remainder is r_k when we carry out the Euclidean algorithm on a and b, then the first zero remainder is $t^{r_k} - 1$ when we carry out the Euclidean

---

[3] The nth Fibonacci number is defined recursively by the formula $F_{n+2} = F_{n+1} + F_n$, where $F_0 = 0$, $F_1 = 1$. Thus the Fibonacci sequence begins 0, 1, 1, 2, 3, 5, 8, ... In a similar way the nth Lucas number is defined recursively by the formula $L_{n+2} = L_{n+1} + L_n$, where $L_0 = 2$, $L_1 = 1$. Thus the Lucas sequence begins 2, 1, 3, 4, 7, 11, 18, ...

algorithm on $t^a - 1$ and $t^b - 1$; so, remembering that in the first part of this proof we found that $r_{k-1} = d$, we have

$$\gcd(t^a - 1, t^b - 1) = t^{r_{k-1}} - 1 = t^d - 1. \qquad \square$$

From Theorem 4.1 we get a consequence that enables us to answer our first question. Consequences of theorems that follow quickly from the statement of the theorem are often called by mathematicians ***corollaries*** of that theorem.

Corollary 4.2 $(t^b - 1) \mid (t^a - 1) \Leftrightarrow b \mid a$.

Proof. We claim that it is obvious, from simple algebra, that if $b \mid a$ then $(t^b - 1) \mid (t^a - 1)$ (this is, in fact, just the case $r_1 = 0$ of the argument above). Conversely, suppose $(t^b - 1) \mid (t^a - 1)$. Then

$$\gcd(t^a - 1, t^b - 1) = t^b - 1,$$

so $t^b - 1 = t^d - 1$, by Theorem 4.1. But it is plain that

$$t^b - 1 = t^d - 1 \Rightarrow b = d. \qquad \square$$

Notice that, as claimed, the condition that $(t^b - 1) \mid (t^a - 1)$ is independent of t.

Corollary 4.3 $(t^n - 1) \mid (t^{n+m} - 1) \Leftrightarrow n \mid m$.
  You shouldn't need us to prove this for you!

Since we now know that, for $\dfrac{t^{n+m} - 1}{t^n - 1}$ to be an integer, it must be true that $n \mid m$, it will be to our advantage, when considering integers of this form, to change notation, writing instead

$$s = \frac{t^{xy} - 1}{t^x - 1}, \qquad (4.3)$$

so that

$$n = x, \qquad m + n = xy. \qquad (4.4)$$

However, we should note that the condition $m \geq 1$, $n \geq 1$ of (4.1) translates into the condition

$$x \geq 1, \qquad y \geq 2 \qquad (4.5)$$

on the integers x, y. We now prove

Table 4.1 *Folding numbers written in base t.*

4	1111	1010101	1001001001	1000100010001
3	111	10101	1001001	100010001
2	11	101	1001	10001
1	1	1	1	1
$\overset{\uparrow}{y} / x \to$	1	2	3	4

Theorem 4.4 *The integer s in (4.3) determines the values of x, y.*

Proof. If we write $\dfrac{t^{xy} - 1}{t^x - 1}$ in base t, then, using the fact that

$$\frac{t^{xy} - 1}{t^x - 1} = t^{x(y-1)} + t^{x(y-2)} + \cdots + t^x + 1$$

we see that

$$s = \frac{t^{xy} - 1}{t^x - 1} \overset{(t)}{=} \underbrace{10 \cdots 0}\ \underbrace{10 \cdots 0} \cdots \underbrace{10 \cdots 0}\ 1, \tag{4.6}$$

where the block $\underbrace{10 \cdots 0}$, of length x, is repeated and there are y 1's. Thus x and y are determined by (4.3). $\square$

We call (x, y) the ***coordinates*** of s and may sometimes, as with points in the plane, even write

$$s = (x, y). \tag{4.7}$$

We write $\mathfrak{F}_t$ for the set of (***integral***) ***folding numbers*** s (***in base t***), given by (4.3), abbreviating simply to ***folding numbers*** if no confusion would occur. Remember the restrictions $x \geq 1$, $y \geq 2$; however, it is sometimes useful to allow $y = 1$. Of course, if $y = 1$, then $s = 1$, independently of the choice of x, so we lose uniqueness there. We may write $\overline{\mathfrak{F}_t}$ for $\mathfrak{F}_t \cup (1)$, that is, $\mathfrak{F}_t$ with the degenerate folding number 1 adjoined. Table 4.1 shows how a table containing values of (x, y), for $1 \leq x, y \leq 4$, would appear in base t. In Table 4.2 we have written the entries in base 10 to display the actual magnitudes of the folding numbers in $\mathfrak{F}_2$.

Formula (4.3) tells us how to recognize folding numbers and (4.4) then gives us the folding procedure to use in order to construct a regular polygon with s sides.

For example, reverting to the base $t = 2$, we observe that

$$341 \overset{(2)}{=} 101010101.$$

Table 4.2 *Folding numbers in base 10.*

y	$2^y - 1$	1	2	3	4	5	6	7	8	9	10	11	12	13	$14 \le x \le 26$
26	67 108 863														
25	33 554 431														
24	16 777 215														
23	8 388 607														
22	4 194 303														
21	2 097 151														
20	1 048 575														
19	524 287														
18	262 143														
17	131 071														
16	65 535														
15	32 767														
14	16 383														
13	8191	8191	22 369 621												
12	4095	4095	5 592 405												
11	2047	2047	1 398 101												
10	1023	1023	349 525												
9	511	511	87 381	19 173 961											
8	255	255	21 845	2 396 745											
7	127	127	5461	299 593	17 895 697										
6	63	63	1365	37 449	1 118 481	34 636 833									
5	31	31	341	4681	69 905	1 082 401	17 043 521								
4	15	15	85	585	4369	33 825	266 305	2 113 665	16 843 009						
3	7	7	21	73	273	1057	4161	16 513	65 793	262 657	1 049 601	4 196 353	16 781 313	67 117 057	
2	3	3	5	9	17	33	65	129	257	513	1025	2049	4097	8193	
1	1	1	1	1	1	1	1	1	1	1	1	1	1	1	
y/x		1	2	3	4	5	6	7	8	9	10	11	12	13	$14 \le x \le 26$

($2^x + 1$)

Thus $x = 2$, $y = 5$, so $n = 2$, $m + n = 10$, and hence $m = 8$; we conclude that the $D^8 U^2$-procedure allows us to construct a regular convex 341-gon. On the other hand

$$11 \overset{(2)}{=} 1011,$$

and hence is not a folding number (in base 2). Thus there is no 2-period folding procedure to fold a regular convex 11-gon.

*4.2 Recognizing rational numbers of the form $\dfrac{t^a - 1}{t^b - 1}$

We know now which integers belong to $\mathfrak{F}_t$ and how to recognize them. What about rational numbers of the form $\dfrac{t^a - 1}{t^b - 1}$? Let us assume that $a > b$ as before. First, we reduce the fraction, and Theorem 4.1 tells us how to do it – we have

$$\frac{t^a - 1}{t^b - 1} = \frac{t^a - 1}{t^d - 1} \bigg/ \frac{t^b - 1}{t^d - 1}, \qquad \text{where} \quad d = \gcd(a, b) \qquad (4.8)$$

and the fraction on the right of (4.8) is reduced. Moreover, $\dfrac{t^a - 1}{t^d - 1} \in \mathfrak{F}_t$ and $\dfrac{t^b - 1}{t^d - 1} \in \overline{\mathfrak{F}_t}$; in fact, $\dfrac{t^b - 1}{t^d - 1} = 1$ precisely when our original number $\dfrac{t^a - 1}{t^b - 1}$ is an integer. Suppose this is not the case.

Let $a = a'd$, $b = b'd$. Then $\gcd(a', b') = 1$ and

$$\frac{t^a - 1}{t^d - 1} = (d, a'), \qquad \frac{t^b - 1}{t^d - 1} = (d, b'). \qquad (4.9)$$

Thus the reduced fraction (on the right of (4.8)) has a folding number (d, a') as numerator and a **prime section** of it, namely (d, b'), as denominator. Here we say that (x, y') is a **prime section** of (x, y) if y' is coprime to y and less than y.

These then are the rational numbers expressible as $\dfrac{t^a - 1}{t^b - 1}$. If we have such a rational number $\dfrac{t^a - 1}{t^b - 1}$, how do we fold tape to produce the star $\left\{ \dfrac{t^a - 1}{t^b - 1} \right\}$-gon? Well, given a folding number (d, a') and a prime section (d, b'), we have $a = a'd$, $b = b'd$, so $m + n = a'd$, $n = b'd$, and $m = (a' - b')d$. Thus, for example, we can fold tape to produce a regular $\left\{ \frac{341}{21} \right\}$-gon since, in base 2,

$$341 \overset{(2)}{=} 101010101$$

$$21 \overset{(2)}{=} 10101.$$

Hence $d=2, a'=5, b'=3$, so $n=b'd=6, m=(a'-b')d=4$, and we conclude that the D^4U^6-folding procedure produces a $\left\{\frac{341}{21}\right\}$-gon; notice that

$$\frac{2^{m+n}-1}{2^n-1}=\frac{2^{10}-1}{2^6-1}=\frac{1023}{63}=\frac{341}{21},$$

as it should!

Suppose now that we really want to produce an 11-gon. In the course of this chapter we will give you two procedures. The first, which we now describe, is simpler, but messier in practice. Although 11 is not a folding number, there are folding numbers having 11 as a factor; for example, 33 is a folding number. If we produce a regular 33-gon, we can easily (in principle!) produce a regular 11-gon simply by visiting consecutively every third vertex of our regular 33-gon.

We will prove a theorem that tells us that every odd number a appears as a factor of some $(x, y) \in \mathfrak{F}_2$; and that we may, in fact, choose x arbitrarily. Which would be the best folding number to choose as a multiple of a? The number-theorist would surely say "the smallest," while the paper-folder would say "that involving the smallest total number of folds" – this number being $m+n$, or xy. Miraculously, these two are the same as we show in Theorem 4.5 below. We call this a miracle because, of course, there is no guarantee that, of two folding numbers, the smaller requires the fewer folds.

Thus

$$1023 = \frac{2^{10}-1}{2^1-1} \quad \text{requires 10 folds,}$$

$$257 = \frac{2^{16}-1}{2^8-1} \quad \text{requires 16 folds.}$$

As usual, we work with an arbitrary base t and consider a fixed but arbitrary integer[4] a coprime to t. We prove

Theorem 4.5 *Every column x of Table* 4.1 *contains a first entry at height $y_0 = y_0(x)$ such that $a \mid (x, y_0)$; and the entries (x, y) in column x such that $a \mid (x, y)$ are precisely those for which y is a multiple of y_0.*

Proof. Now t is coprime to a and to $t^x - 1$, hence to $a(t^x - 1)$. Thus it follows that there exists a positive integer z_0 such that[5]

$$t^z \equiv 1 \bmod a(t^x - 1),$$

[4] We may assume $a > 1$. This use of the symbol "a" is, of course, not to be confused with previous uses of the letter a.
[5] By definition $A \equiv B \bmod m$ means $m \mid A - B$; for example $32 \equiv 2 \bmod 5$, since $5 \mid 32 - 2$.

if and only if z is a multiple of z_0. It then follows that $t^x - 1 \mid t^{z_0} - 1$ so that, by Corollary 4.3, $x \mid z_0$, and we may write $z_0 = x y_0$. Hence any multiple of z_0 is of the form xy, where y is a multiple of y_0. We conclude that

$$a \left| \frac{t^{xy} - 1}{t^x - 1} \Leftrightarrow y \text{ is a multiple of } y_0, \right.$$

as claimed.   □

Let $h(a) = y_0(1)$; we call $h(a)$ the **height** of a. It is the height, in Table 4.1, of the first element of the form $(1, y)$ which is divisible by a.

Proposition 4.6 *Every folding number (x, y), with base t such that $a \mid (x, y)$ satisfies $h(a) \mid xy$.*

Proof. Because $a \left| \dfrac{t^{xy} - 1}{t^x - 1} \right.$, it follows that $a \left| \dfrac{t^{xy} - 1}{t - 1} \right.$, and thus $h(a) \mid xy$ by Theorem 4.5.   □

We come now to our key result:

Theorem 4.7 *The smallest folding number s having a as a factor is of the form (x, y) with*

$$xy = h(a).$$

Proof. Suppose not. Then, if $s = (x, y)$ is the smallest folding number having a as a factor, we have (by Proposition 4.6) $h \mid xy$, but $h \neq xy$ (where $h = h(a)$). Thus $xy \geq 2h$. Now remember from (4.5) that $y \geq 2$. Thus (you should check each step in the line below)

$$s = \frac{t^{xy} - 1}{t^x - 1} > \frac{t^{xy}}{t^x} = t^{x(y-1)} \geq t^{\frac{1}{2}xy} \geq t^h > \frac{t^h - 1}{t - 1}. \qquad (4.10)$$

But, by the definition of h, the folding number $\dfrac{t^h - 1}{t - 1}$ is divisible by a, so that (4.10) contradicts the minimality of s. We are thus forced to conclude that $h = xy$, as claimed.   □

We draw attention to an immediate and remarkable consequence of Theorem 4.7. We think of $s = (x, y)$ *as a function of t* and prove

Corollary 4.8 *Let $s = (x, y)$, $s' = (x', y')$. Then if, for some t, $s \mid s'$, it follows that $xy \mid x'y'$.*

Proof. We fix a value of t such that $s \mid s'$. We let s play the role of a in Proposition 4.6. Then s is obviously the smallest folding number divisible by s, so that $h(s) = xy$. But Proposition 4.6 guarantees that $h(s) \mid x'y'$.   □

This result is remarkable since the conclusion is independent of t and only requires one value of t such that $(x, y) \mid (x', y')$. As we will see, there may well be only one such value of t. Notice also that our proof of this striking result of pure number theory was inspired by very practical considerations of efficient paper-folding.

We also see in the statement of Corollary 4.8 the great advantage of well-conceived generalization. If we had stuck to base 2 we would have had a vastly inferior result in place of Corollary 4.8. In fact, we are now able to raise a question which is natural but would have been quite beyond us. Notice that, with this question, we leave the realm of geometry and firmly enter that of number theory; but it is the mathematics itself which has compelled us to make this move. Our question is:

> **Given $s = (x, y)$, $s' = (x', y')$, for which t is it true that $s \mid s'$?**

We will describe the answer to this question, which may take a form that, at first glance, will disappoint you.

Of course, we may suppose that $xy \mid x'y'$; otherwise, as Corollary 4.8 tells us, the answer is that $s \mid s'$ for no value of t. Now let $\gcd(xy, x') = d$, so that $xy = ad$, $x' = bd$, with $\gcd(a, b) = 1$. Since $xy \mid x'y'$, it follows that $ad \mid bdy'$, so that $a \mid by'$. But $\gcd(a, b) = 1$, so that $a \mid y'$. This prepares the way for our answer, which is contained in the following theorem.

Theorem 4.9 *Suppose* $xy \mid x'y'$ *and let* $d = \gcd(xy, x')$ *and* $h = \gcd(x, x')$. *Then*

$$\frac{t^{xy} - 1}{t^x - 1} \left| \frac{t^{x'y'} - 1}{t^{x'} - 1} \right. \Leftrightarrow \frac{t^d - 1}{t^h - 1} \left| \frac{y'}{a} \right. .$$

This is proved in [27] and [30] and we will not repeat the proof here. Rather we want to show why the theorem is very useful. First notice that

$$h = \gcd(x, x') = \gcd(x, xy, x') = \gcd(x, d).$$

Next, notice that $\dfrac{y'}{a}$ is independent of t. There are thus two possibilities:

Case 1

$h = d$, that is, $d \mid x$. Then $\dfrac{t^d - 1}{t^h - 1} = 1$, whatever the value of t, and, of course, $1 \mid \dfrac{y'}{a}$. Thus, in this case,

$$s \mid s' \quad \text{for all } t.$$

An example of this case is provided by

$$s = \frac{t^{45} - 1}{t^9 - 1}, \qquad s' = \frac{t^{90} - 1}{t^6 - 1}.$$

Then $d = \gcd(45, 6) = 3$ and $x = 9$ so $d \mid x$. We may easily verify the truth of our claim in this case since

$$\frac{s'}{s} = \frac{(t^{90} - 1)(t^9 - 1)}{(t^{45} - 1)(t^6 - 1)} = \frac{(t^{45} + 1)(t^9 - 1)}{(t^3 + 1)(t^3 - 1)},$$

which is obviously a polynomial in t with integer coefficients.

Case 2

$h \neq d$, that is, $d \nmid x$. Then $h < d$, so that $\dfrac{t^d - 1}{t^h - 1}$ is a polynomial in t of positive degree $d - h$ and leading term t^{d-h}. It follows that $\dfrac{t^d - 1}{t^h - 1}$ tends to infinity with t and can thus be less than or equal to $\frac{y'}{a}$ (which, as we have already pointed out, is a constant independent of t) for only finitely many integer values of t. Thus, in this case,

$$s \mid s' \quad \text{for only finitely many values of } t.$$

We immediately infer the striking result:

Theorem 4.10 *If $s \mid s'$ for infinitely many t, then $s \mid s'$ for all t.*

In fact, we have a stronger statement. Remember that we may think of s and s' as polynomials in t with integer coefficients and leading coefficient 1.

Theorem 4.11 *The following statements are equivalent:*

(i) *$s \mid s'$ for all t;*
(ii) *$s \mid s'$ for infinitely many t;*
(iii) *$s \mid s'$ as polynomials with integer coefficients.*

Proof. We know that (ii) ⇒ (i), and it is obvious that (iii) ⇒ (i) (see our analysis of the example given in Case 1). Thus it remains to show that (ii) ⇒ (iii) for, of course, (i) ⇒ (ii). In fact, we prove a more general result which is no harder.

Let $f(t), g(t)$ be two polynomials with integer coefficients, and let g have leading coefficient 1. Suppose that $g(t)$ divides $f(t)$ for infinitely many (positive integer) values of t. *Then $g(t)$ divides $f(t)$ as a polynomial.*

To see this, first divide $f(t)$ by $g(t)$, getting a quotient $q(t)$ and a remainder $r(t)$; both of these will also be polynomials with integer coefficients. We want to show that $r(t)$ is the zero polynomial. If not it is a polynomial of degree less than that of $g(t)$, and

$$f(t) = q(t)g(t) + r(t). \tag{4.11}$$

Suppose that $g(t)$ divides $f(t)$ for the increasing sequence of integers $t = t_1, t_2, t_3, \ldots$ Since $r(t)$ has only finitely many zeros, we may drop these from the list (if they occurred) and still have an infinite sequence. Then (4.11) shows that $g(t)$ divides $r(t)$ for these same values of t. Now degree $g(t) >$ degree $r(t)$. This implies that $\dfrac{r(t)}{g(t)}$ tends to 0 as t tends to infinity, so that, for t sufficiently large,

$$\left| \frac{r(t)}{g(t)} \right| < 1.$$

But, however large we take t, there is some t_N in our sequence of integers which is larger than t and for which $g(t_N)$ divides $r(t_N)$. We are in a hopeless contradiction – how can $g(t_N)$ divide $r(t_N)$ as integers when the ratio

$$\left| \frac{r(t_N)}{g(t_N)} \right|$$

is less than 1 and $r(t_N) \neq 0$? Thus $r(t)$ must be the zero polynomial, as our statement claimed. □

*4.3 Numerical examples and why $3 \times 7 = 21$ is a very special number fact

We give two examples, treated in some detail. The first, Example 4.1, is beyond the scope of a modern computer, if tackled head-on. The second, Example 4.2, exhibits a surprising feature of the well-known number fact $3 \times 7 = 21$.

Example 4.1 Let $s = (2, 4) = \dfrac{t^8 - 1}{t^2 - 1}$, $s' = (4, 40) = \dfrac{t^{160} - 1}{t^4 - 1}$. Then $d = \gcd(8, 4) = 4$, $h = \gcd(2, 4) = 2$. Moreover, $8 = 4a$, so $a = 2$, and $y' = 40$.

Thus

$$s \mid s' \Leftrightarrow \frac{t^4 - 1}{t^2 - 1} \mid 20 \Leftrightarrow t^2 + 1 \mid 20.$$

It is easy to see that this holds for precisely $t = 2, 3$ (remember that $t \geq 2$). We are in Case 2 and $s \mid s'$ only when $t = 2, 3$.

Example 4.2 We consider the innocent statement $21 = 7 \times 3$. Now 21 and 7 are folding numbers,

$$21 = \frac{2^6 - 1}{2^2 - 1}, \quad 7 = \frac{2^3 - 1}{2^1 - 1}, \qquad \text{or} \qquad 21 = (2, 3), \quad 7 = (1, 3).$$

We look at a general base t and apply Theorem 4.10. It is easy to see that we are in Case 1, so that

$$\frac{t^3 - 1}{t - 1} \Bigm| \frac{t^6 - 1}{t^2 - 1} \qquad \text{for } all \text{ bases } t.$$

Thus, for instance

$$(t = 3) \quad 13 \mid 91$$
$$(t = 4) \quad 21 \mid 273$$
$$(t = 5) \quad 31 \mid 651 \ldots$$

If we write 7 and 21 in base 2 we get 111 and 10101; and we may restate our conclusion by saying that $111 \mid 10101$, interpreted in *any* base. Thus the statement $7 \mid 21$ is seen as just one case of a completely general fact. Of course, by elementary algebra we can see that

$$\frac{t^3 - 1}{t - 1} \text{ must divide } \frac{t^6 - 1}{t^2 - 1} \text{ as polynomials;}$$

in fact,

$$\frac{(t^6 - 1)(t - 1)}{(t^3 - 1)(t^2 - 1)} = t^2 - t + 1.$$

Now consider 21 and 3; 3 is also a folding number,

$$3 = \frac{2^2 - 1}{2^1 - 1} = (1, 2).$$

We look at a general base t and ask when does

$$\frac{t^2 - 1}{t - 1} \Bigm| \frac{t^6 - 1}{t^2 - 1}?$$

Applying Theorem 4.10, we conclude that

$$\frac{t^2 - 1}{t - 1} \left| \frac{t^6 - 1}{t^2 - 1} \right. \Leftrightarrow \frac{t^2 - 1}{t - 1} \left| 3 \Leftrightarrow (t + 1) \right| 3.$$

Since $t \geq 2$, we conclude that

$$\frac{t^2 - 1}{t - 1} \left| \frac{t^6 - 1}{t^2 - 1} \right. \qquad only \text{ when } t = 2.$$

We are in Case 2 with a vengeance! We now know that $11 \mid 10101$ in base 2 *and in no other base*. Thus the statement $3 \mid 21$ is seen as a highly singular phenomenon! Or, as we sometimes like to say,

> 7 divides 21 always
> but 3 divides 21 almost never!

We close this section with the remark that there is one further generalization which we might have introduced at almost no cost in complication. Instead of looking at numbers

$$s(x, y) = \frac{t^{xy} - 1}{t^x - 1},$$

we could have looked at numbers

$$r(x, y) = \frac{t^{xy} - u^{xy}}{t^x - u^x},$$

where t, u are coprime (we may assume $t > u$). All our results, suitably modified, hold in this generality – and, of course, then become even more striking. You can find the details in [32].

You may wish to experiment and try to *guess* some theorems and results for numbers of the form

$$r(x, y) = \frac{t^{xy} - u^{xy}}{t^x - u^x}, \qquad \text{where } t, u \text{ are coprime and } t > u.$$

For example, verify with a hand calculator (comparing the results with Example 4.2) that

(a) $\dfrac{5^3 - 3^3}{5 - 3} \left| \dfrac{5^6 - 3^6}{5^2 - 3^2} \right.$ but $\dfrac{5^2 - 3^2}{5 - 3} \nmid \dfrac{5^6 - 3^6}{5^2 - 3^2};$

(b) $\dfrac{7^3 - 2^3}{7 - 2} \left| \dfrac{7^6 - 2^6}{7^2 - 2^2} \right.$ but $\dfrac{7^2 - 2^2}{7 - 2} \nmid \dfrac{7^6 - 2^6}{7^2 - 2^2}.$

This can clearly be seen by observing that the quotient when $\dfrac{t^6 - u^6}{t^2 - u^2}$ is divided by $\dfrac{t^3 - u^3}{t - u}$, as a polynomial in t and u, is $t^2 - tu + u^2$.

4.4 A number trick and two mathematical tidbits

We insert here three attractive pieces of mathematics related to the set of folding numbers $\mathfrak{F}_t$.

A number trick

A nice number trick which you can try on your friends is to exploit the identity

$$(x, yz) = (xy, z)(x, y). \tag{4.12}$$

Of course, (4.12) simply reflects the obvious identity

$$\frac{t^{xyz} - 1}{t^x - 1} = \frac{t^{xyz} - 1}{t^{xy} - 1} \cdot \frac{t^{xy} - 1}{t^x - 1};$$

but it is very impressive if you happen to be at a party where a table of $\mathfrak{F}_2$ (see Table 4.2) is prominently displayed. What you do is (appearing to think very hard) divine that the entry in position $(3, 6)$, that is, 37 449, is the product of the entry in position $(9, 2)$, that is, 513, and the entry in position $(3, 3)$, that is, 73. Try some other examples for yourself.

Tidbit 1

We proved (Theorem 4.1 of this chapter) that

$$\gcd(t^a - 1, t^b - 1) = t^d - 1, \qquad \text{where} \quad d = \gcd(a, b).$$

It is natural to ask whether the similar relationship holds true for the lcm,[6] that is, whether

$$\mathrm{lcm}(t^a - 1, t^b - 1) = t^\ell - 1, \qquad \text{where} \quad \ell = \mathrm{lcm}(a, b). \tag{4.13}$$

Now there is an obvious case in which (4.13) does hold. If $b \mid a$, then $t^b - 1 \mid t^a - 1$, and $a = \mathrm{lcm}(a, b)$, so

$$\mathrm{lcm}(t^a - 1, t^b - 1) = t^a - 1 = t^\ell - 1.$$

---

[6] lcm, of course, stands for "least common multiple."

Likewise (4.13) holds if $a \mid b$. The remarkable conclusion which we can draw, however, is that (4.13) holds *only* in these two very special cases!

The argument runs as follows. We first prove a lemma, that is, a helpful result which will enable us to answer the question.

Lemma 4.12 *The expression of a rational number different from 1 as*

$$\frac{t^p - 1}{t^q - 1}$$

(if it exists) is unique.

Proof. We must show that, if

$$\frac{t^p - 1}{t^q - 1} = \frac{t^r - 1}{t^s - 1},$$

then either $p = q, r = s$ or $p = r, q = s$. Now if

$$\frac{t^p - 1}{t^q - 1} = \frac{t^r - 1}{t^s - 1},$$

then

$$t^{p+s} - t^p - t^s = t^{r+q} - t^r - t^q.$$

Suppose, as we may, that q is the *least* of the numbers p, q, r, s. Then, dividing by t^q, we get an obvious contradiction (remember $t \geq 2$) if q is *uniquely* the least, so we must suppose it is not. Thus $q = p$, $q = s$, or $q = r$. We are happy with $q = p, q = s$, since $q = p$ implies $r = s$ and $q = s$ implies $p = r$. But if $q = r$ and $q \neq p, q \neq s$, we have another obvious contradiction, since then

$$\frac{t^p - 1}{t^q - 1} > 1, \qquad \frac{t^r - 1}{t^s - 1} < 1.$$

This proves the lemma. □

Now suppose that (4.13) holds. Then, since for any two numbers m, n,

$$mn = \gcd(m, n)\text{lcm}(m, n) \quad \text{(can you see why this is true?)},$$

we have, by Theorem 4.1,

$$(t^a - 1)(t^b - 1) = (t^d - 1)(t^\ell - 1)$$

so that

$$\frac{t^a - 1}{t^d - 1} = \frac{t^\ell - 1}{t^b - 1}.$$

Lemma 4.12 now comes into play to tell us that

$$d = a, \quad b = \ell, \quad \text{that is,} \quad a \mid b$$

or

$$d = b, \quad a = \ell, \quad \text{that is,} \quad b \mid a.$$

It is amusing to see that (4.13) cannot hold precisely because Theorem 4.1 *does* hold.

Tidbit 2

We have pointed out in the previous section the important fact that, given any number a prime to t, there are numbers in $\mathfrak{F}_t$ having a as a factor; and we have further emphasized the importance of finding the smallest such number s in $\mathfrak{F}_t$. Is there a nice algorithm for finding s? The answer is yes!

First, we must determine $h(a)$, the *height* of a (see the definition following the proof of Theorem 4.5). Now by a generalization of Fermat's Little Theorem we know (see [24] Chapter 2) that

$$t^{\Phi(a(t-1))} \equiv 1 \bmod a(t-1),$$

where Φ is the Euler totient function; that is $\Phi(n)$ is the number of positive integers less than or equal to n that are coprime (relatively prime) to n.

Thus

$$h(a) \mid \Phi(a(t-1)).$$

In fact, we are looking for the *smallest* factor h of $\Phi(a(t-1))$ such that

$$t^h \equiv 1 \bmod a(t-1).$$

Suppose

$$\Phi(a(t-1)) = p_1^{m_1} p_2^{m_2} \cdots p_k^{m_k}$$

is a factorization of $\Phi(a(t-1))$ as a product of primes. We test

$$h_1 = p_1^{m_1-1} p_2^{m_2} \cdots p_k^{m_k}$$

to see if $t^{h_1} \equiv 1 \bmod a(t-1)$. If so, we test

$$h_2 = p_1^{m_1-2} p_2^{m_2} \cdots p_k^{m_k}$$

and continue to lower the exponent of p_1 by 1 until *either* we reach

$$h_i = p_1^{m_1-i} p_2^{m_2} \cdots p_k^{m_k}$$

such that $t^{h_i} \not\equiv 1 \bmod a(t-1)$ or the exponent reaches zero, and still $t^{h_i} \equiv 1 \bmod a(t-1)$. In the former case we fix the exponent of p_1 at $m_1 - i + 1 (= n_1)$ and start again with h_{i-1} which we now call

$$h_{i+1} = p_1^{n_1} p_2^{m_2} \cdots {}_k^{m_k}.$$

Now, however, we operate on the exponent of p_2 just as we did above for the exponent of p_1. In the latter case we eliminate p_1 and start with

$$h_i = p_2^{m_2} \cdots p_k^{m_k},$$

now operating on the exponent of p_2 just as we did above for the exponent of p_1. Then, after handling p_2, we handle $p_3, p_4, \ldots, p_k$. At the last step q for which

$$h_q \equiv 1 \bmod a(t-1),$$

we will have found $h(a)$. Notice that we can choose the order of the prime factors ourselves.

Let us give an example; we take $t = 3$, $a = 14$. Now

$$\Phi(28) = \Phi(4)\Phi(7) = 2 \times 6 = 12,$$

so we know that $3^{12} \equiv 1 \bmod 28$. Since $12 = 2^2 \cdot 3$, we first try $h_1 = 2 \cdot 3$. Then $3^6 - 1 = 728$, which is certainly divisible by 28. Next we try $h_2 = 3$, but $3^3 \not\equiv 1 \bmod 28$. Thus $h_3 = h_1 = 2 \cdot 3 = 6$ and we must finally try $h_4 = 2$ but again $3^2 \not\equiv 1 \bmod 28$, so $h(14) = h_3 = h_1 = 6$.

Now we know how to calculate $h(a)$, we also know that s will be the smallest integer expressible as

$$\frac{t^{h(a)} - 1}{t^x - 1}$$

which is a multiple of a. (Certainly, if $x = 1$, then $\dfrac{t^{h(a)} - 1}{t^x - 1}$ is a multiple of a.) Thus an easy algorithm is to examine the proper factors of $h(a)$ in descending order, stopping the first time we find such a factor x that $\dfrac{t^{h(a)} - 1}{t^x - 1}$ is a multiple of a.

Let us return to our example $t = 3$, $a = 14$. We found $h(a) = 6$ so we consider numbers (in $\mathfrak{F}_3$) of the form $\dfrac{3^6 - 1}{3^x - 1}$ with x a proper factor of 6. We first try $x = 3$.

Then $\dfrac{3^6 - 1}{3^3 - 1} = 3^3 + 1 = 28$ which is a multiple of 14, so we have found the required s.

As a second example, take $t = 2$, $a = 21$. Then $\Phi(a) = 12$ and we quickly find that $h(a) = 6$. We consider numbers of the form $\dfrac{2^6 - 1}{2^x - 1}$, with x a proper factor of 6. With $x = 3$, we get 9 which is no good. With $x = 2$ we get 21, which works. Indeed, in this case, our algorithm has obtained for us the fact we already know, namely, that 21 is itself a folding number.

A particularly easy case is that in which a is prime. For let x be the *largest* proper factor of $h(a)$. Then we claim that

$$a \left| \frac{t^{h(a)} - 1}{t^x - 1} \right.,$$

so that

$$s = \frac{t^{h(a)} - 1}{t^x - 1}.$$

For

$$a \left| \frac{t^{h(a)} - 1}{t^x - 1} \cdot \frac{t^x - 1}{t - 1} \right.$$

but

$$a \not\left| \frac{t^x - 1}{t - 1} \right., \quad \text{by the minimality of } h(a).$$

At this point we quote a well-known theorem of elementary number theory; namely, "if p is prime and $p \mid k\ell$, then $p \mid k$ or $p \mid \ell$." Thus, since a is prime, it follows that

$$a \left| \frac{t^{h(a)} - 1}{t^x - 1} \right..$$

Let us apply this to the case $t = 2$, $a = 13$. Then

$$\Phi(a) = 12 = 2^2 \cdot 3.$$

We try $h_1 = 2 \cdot 3 = 6$, but $2^6 \not\equiv 1 \bmod 13$; we try $h_2 = 2^2$, but $2^4 \not\equiv 1 \bmod 13$. Thus $h(a) = 12$. It follows that the folding number s we require is given by

$$s = \frac{2^{12} - 1}{2^6 - 1} = 2^6 + 1 = 65.$$

5

The polyhedron thread – Building some polyhedra and defining a regular polyhedron

5.1 An intuitive approach to polyhedra

In Chapter 3 we looked at polygons in the plane, and in particular, we studied, there and in Chapter 2, how to construct some regular convex polygons. A natural extension of this idea in 3 dimensions is to study how to construct *polyhedra*, which are, in an obvious way, the 3-dimensional analogs of polygons. For example, just as a connected polygon divides the plane into two regions (the inside and the outside), a connected polyhedron divides space into two regions (again, the inside and outside). A polygon consists of straight (uncurved) sides, that is, parts of lines, whereas a polyhedron consists of flat (uncurved) faces, that is, parts of planes. A rectangular box is an example of a polyhedron, but a cylindrical can is not because its boundary does not consist entirely of flat faces.

The formal study of polyhedra is very rich and intellectually rewarding, but we will restrain ourselves and postpone our general discussions of the mathematics of polyhedra, which is by no means exhaustive, until Chapters 9, 14, and 15. Here, and in Chapter 8, we study polyhedra from the practical, constructive point of view. We think this is the appropriate order of events. If you choose to build the models we describe, you will discover that making them is a vivid educational experience, particularly if you have never built polyhedral models before. Sometimes it will seem almost magical when the faces finally all fit together – and sometimes the final shape obtained is surprisingly beautiful. Of course, it can be exasperating if the pieces don't fit together correctly, but we've tried to spare you this unpleasant experience by including in our instructions more information than is actually needed to construct the polyhedra, some of it of a very practical non-mathematical nature.

Nevertheless, most people find constructing polyhedra so exciting and absorbing that it would be quite pointless to try to do *anything* else at the same time. Teachers, especially, should be aware of this and should let their students enjoy the experience of constructing the polyhedra without other distractions. Don't be too impatient, wait, and discuss the mathematical properties of the models *after* some have been constructed.

In the next section we describe a classical method for constructing a special class of polyhedra, called **convex deltahedra** from what are called **net diagrams**, and then we make some suggestions for how you can use what you learned in Chapter 2 to produce the nets. When you have constructed the models, you will be prepared to read Section 5.3, where we ask and answer the questions, What is a polyhedron? and What is a *regular* convex polyhedron?

But let us get on with constructing some polyhedra.

5.2 Constructing polyhedra from nets

Required materials

- $8\frac{1}{2} \times 11$ inch (21×27 cm, approximately $A4$ size), or bigger, sheets of heavy construction paper (colored paper is nice)
- Pen or pencil
- White glue, or glue stick
- Scissors

Optional materials

- Gummed mailing tape about 2 inches (5 cm) wide (the heaviest you can find)
- Sponge
- Bowl of water
- Hand towel (or rag)
- Books
- Colored paper, acrylic paint, or glitter

The idea is simple. Think of taking a polyhedron, such as a cube, that is made from paper and slitting apart some of the edges so that all of the faces of the polyhedron lie flat and the whole thing still remains in one piece. The resulting configuration (of which, for any given polyhedron, there are many possibilities) is called a **net** for the polyhedron. Now, since our object is really to go the other way, that is, to *construct* the polyhedron from a net, it is necessary to add some tabs to the net, producing what we call a **complete net**, so that we can glue the edges together. It is an interesting and useful fact that for any net of a polyhedron, if tabs are attached to alternate sides of the boundary to form a complete net, then it will always be possible to assemble the polyhedron by using those tabs to join appropriate faces. Notice that we said *appropriate* faces. We were very careful about this because, as you may easily verify – in the case of the triangular dipyramid, for example – it is sometimes possible to join the faces of the complete net by means of the tabs in such a way that you don't get the desired polyhedron – or, in fact, any polyhedron at all! Where we think it would be helpful, we have added arrows to the complete net to indicate that the two sides at the beginning and end of the arrow should be joined to form an edge of the polyhedron.

Figure 5.1 shows complete nets for a special class of polyhedra called ***convex deltahedra***. *A **deltahedron*** is a polyhedron, all of whose faces are triangles (we will explain the word *convex* in the next section). An illustration of the completed model and its name is shown next to each complete net so that you will have an idea of how the completed polyhedron should look.

It is interesting that there doesn't seem to be any universal agreement about the naming of the convex deltahedra. The 8 convex deltahedra are part of what are known as the Johnson solids;[1] the tetrahedron, octahedron, and icosahedron don't have any identifying Johnson numbers, but the remaining ones bear the identifying numbers, J12 (6 faces), J13 (10 faces), J17 (16 faces), J51 (14 faces), and J84 (12 faces), and are referred to by Johnson as the triangular dipyramid, pentagonal dipyramid, geoelongated square dipyramid, triaugmented triangular prism, and snub disphenoid, respectively. According to Cromwell [10] the 12-faced deltahedron is sometimes called the *Siamese dodecahedron* – a name coined by H. S. M. Coxeter. However colorful these various names may be, we prefer to use the names that tell you how many faces are on the surface of the polyhedron. Cundy and Rollet [12] and Pugh [72] use names similar to the ones we use (except they use "hecca" for six, instead of "hexa"). Other authors, Holden [49] and Beck, Bleicher and Crowe [3], avoid the controversy by discussing the eight deltahedra without giving names to the ones that aren't Platonic solids.

General instructions

1. Make a copy on construction paper of the net diagram of the complete net for the desired polyhedron shown in Figure 5.1.
2. Cut out the complete net along the boundary.
3. Carefully *valley* fold on each of the fold lines shown by the dash-dotted lines (these lines will eventually be *inside* the model). You may need to use a pen or pencil, with a ruler, to gently score the lines so that they will fold neatly.
4. Glue the tabs in place on the *outside* of the model.
5. If you wish to have a more decorative model you can cut out colored triangles and glue them on the faces, paint the faces with a suitable paint, or spread glue on the faces and sprinkle glitter on them. In a word, be creative! (optional).

---

[1] A Johnson solid is a convex polyhedron having regular faces and equal edge lengths (with the exception of the completely regular Platonic solids, the "semiregular" Archimedean solids, and the two infinite families of prisms and antiprisms).

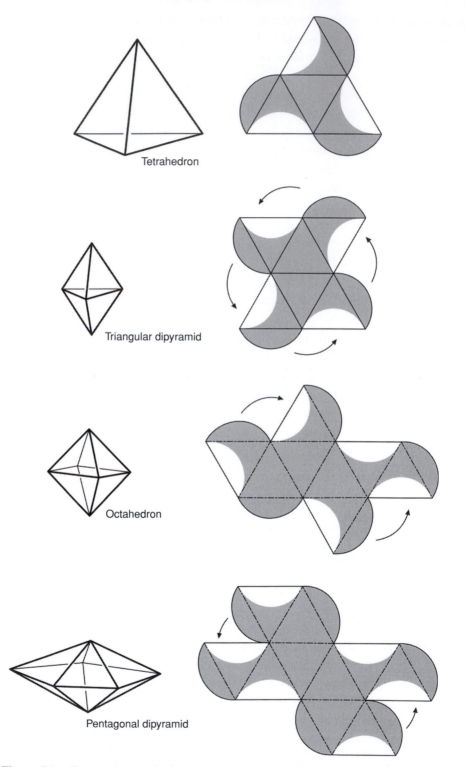

Figure 5.1   Convex deltahedra based on patterns from Pedersen and Pedersen [64].

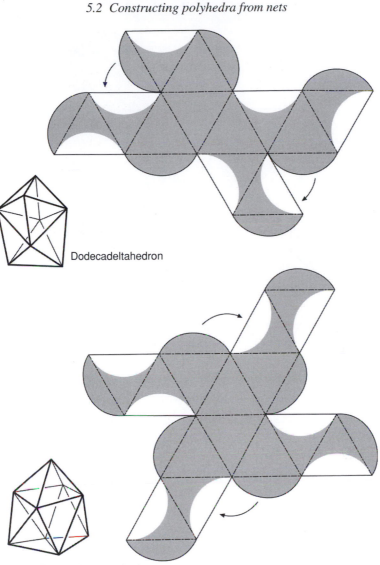

Dodecadeltahedron

Tetracaidecadeltahedron

Figure 5.1   (*cont.*)

General instructions (alternative)

1. Select the polyhedron you want to make.
2. Count the number of faces and tabs.
3. Fold a strip of heavy gummed tape by the $U^1 D^1$-procedure until you have about twice as many triangles as there are faces and tabs on the model you want to make. (Keep any extra triangles for the next model.)
4. Cut and glue portions of the $U^1 D^1$-tape together to make a duplicate of the desired pattern piece (this is where you use the sponge and the bowl, into which you have put

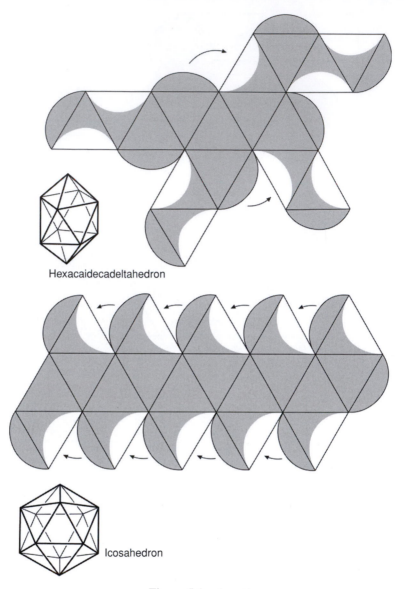

Hexacaidecadeltahedron

Icosahedron

Figure 5.1   (*cont.*)

some water). Do this so that the gummed side will be on the *inside* of the finished model. Here it is particularly easy to make tabs in the shape of equilateral triangles; for ease of construction, mark them so that you will know they are tabs and not faces.

5. Place heavy books on the pattern piece and tabs while they dry. This keeps the faces flat.

6. Valley-fold on all the crease lines that will become edges on the finished model. (This means fold so that the pattern piece will, by itself, begin to "curl" into the shape of the finished model, with the gummed portion *inside* the model.)

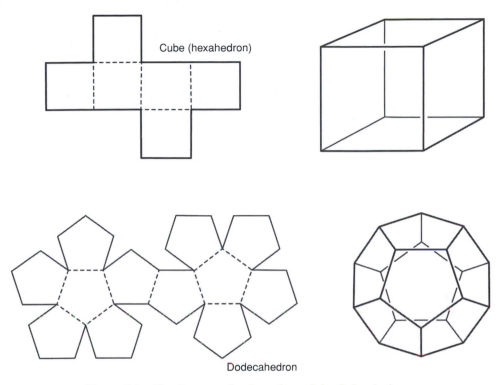

Figure 5.2   Net diagrams for the cube and the dodecahedron.

7. Glue the tabs in place, paying attention to the arrows, if there are any on the net diagram.
8. Color the faces (optional).

Notice that the triangular tabs add strength to the model and, since they cover the entire face, their function is undetectable. Notice, also, that with this method all of the gluing is done on the *outside* of the model. We think this is a distinct improvement over those instructions that tell you to glue the tabs *inside* the polyhedron. (We always find that the last tab is almost impossible to do neatly, if it can be done at all.)

Notice that the idea of constructing polyhedra from nets works just as well if the faces are regular squares or pentagons. In Figure 5.2 we give you the **nets** for the hexahedron, or cube, and the regular pentagonal dodecahedron (which we will refer to simply as the "dodecahedron"). We leave the modification of the traditional general instructions to you (see Figure 5.2). Since the tabs are not shown in Figure 5.2 you will need to supply them. In the case of the cube, you proceed by first making one square cardboard pattern piece and drawing the basic net as shown; then adding the tabs to alternate sides of its boundary to produce a completed net. You

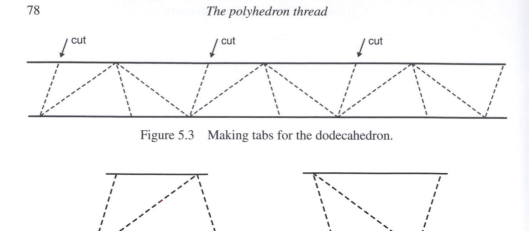

Figure 5.3   Making tabs for the dodecahedron.

Figure 5.4   Two versions of a short section of U^2D^2-tape.

may wish to use tabs that are themselves square, since they will then reinforce the construction and leave no trace of their function.

In a similar way, if you begin with one cardboard pentagon (you can produce the pattern for it by the U^2D^2-folding procedure discussed in Section 3.4), you can draw the basic net for the dodecahedron and add the tabs to alternate sides of the boundary to form the completed pattern piece. Here, it is not possible for all the tabs to be in the shape of a regular pentagon without overlapping each other. However, if the size of the pentagon you used to make your net is the same as that of the pentagon formed by folding along short lines of the U^2D^2-tape, then you can use more of this same U^2D^2-tape to cut out tabs to glue onto the net diagram (using a sponge and water). Simply cut along every other short line as shown in Figure 5.3, to get the tabs. Note that these tabs won't completely cover the faces on either side of the edge to which they are attached, but the geometry will make it plain where they should be glued.

Alternative construction

It is also possible to construct a net from a single U^2D^2-strip of tape. First think of a *short section* of the U^2D^2-tape as shown in Figure 5.4.

Now take a strip of tape containing 30 consecutive short sections and fold it along certain short lines, gluing all the overlapping portions, so that the result looks exactly like the basic net shown in Figure 5.5 (a small sponge and a bowl of water are handy – with a hand towel or rag to wipe up the excess water). *Caution*: Note that the pentagons at the top and bottom of this net are formed from *five* short sections (zigzagging back and forth to form a very sturdy pentagon). Once the net

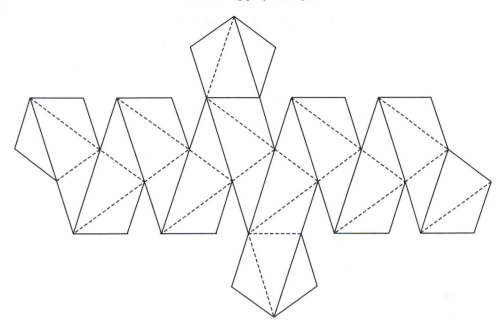

Figure 5.5   A basic net for the dodecahedron.

is prepared, tabs can be made, as before, by cutting the $U^2 D^2$-tape on every other short line. Remember to attach the tabs to alternate sides of the net and recognize that, at this point, there will be lots of overlapping portions of paper.

Practical hints

1. Be certain to glue all of the overlapping portions in the basic net.
2. Let the net, with its tabs attached, dry underneath some heavy books so that the faces will be as flat as possible.
3. Crease all of the edge lines firmly so that the edges will be sharp and straight.
4. Where tabs are attached, crease the edge firmly before gluing the other half of the tab to the model.
5. Be patient; don't try to go too fast. Wait until each tab is stable before gluing the next one.

Now that you have constructed some polyhedra, you should get to know them. Play with them, admire them, and notice how many faces they have, what they look like from various directions, and how they *feel*. Do you notice anything about them when you close your eyes and hold them that you didn't notice by simply *looking* at them? Experiment by holding them in front of a light and looking at the shadows they cast. How many of them can make hexagonal shadows? Try classifying them according to certain properties.

5.3 What is a regular polyhedron?

Before we attempt to answer this question we should be more precise as to what we mean by a polyhedron. Let us begin by looking again at polygons. We distinguish between a *polygon* (or polygonal path) and a *polygonal region*. A polygon consists of edges hinged together at vertices and does not contain its interior (recall that it was the top edge of the tape in Chapters 2 and 3 that formed the FAT polygons). Similarly, we say that a **polyhedron** consists of faces hinged together at edges, and it does not contain its interior. Thus, strictly speaking, a polyhedron is a *surface* and not a *solid*, although it is sometimes loosely referred to as a solid (for example, the Platonic solids, which we will discuss shortly).

Now, just as each vertex of a polygon is an endpoint of exactly two edges, so is each edge of a polyhedron the side of exactly two faces. Thus a polyhedron should be regarded as a collection of faces, each of which is a polygonal region; and two intersecting faces meet precisely in a common side of each, or in just one vertex that is a common vertex of each face. For examples look at the figures next to the net diagrams of Figure 5.1.

A polygon is *connected*, meaning that it is all in one piece; the polyhedra we consider are also connected in this sense. Further, just as we lay particular emphasis on *convex* polygons, so too we confine our attention here to *convex* polyhedra. We define a polyhedron to be *convex* if, given any two points P and Q of the region bounded by the polyhedron, every point of the straight-line segment PQ belongs to that region. Alternatively, we may say that if P and Q are any two points of the polyhedron, the straight-line segment PQ consists of points of the polyhedron or its interior.

So in this section we will be discussing convex connected polyhedra as here defined. We emphasize that this is a very restrictive definition. In particular, the restriction to convex polyhedra is far more significant than the corresponding restriction to convex polygons. Without this restriction, it is not even meaningful to talk of "the region bounded by the polyhedron." You may think of the polyhedra we wish to discuss as being obtained from a spherical surface made of some rubbery material by "pushing and pulling" it around until it consists of flat polygons and faces as described.

It is customary to name polyhedra in a manner similar to the way we named polygons – that is, just as we incorporated the number of sides of a polygon into its name (thus a *pentagon* has 5 sides), we now incorporate the number of faces a polyhedron has into its name.[2] In Table 5.1, we list the names of some of the

---

[2] It is common, in ordinary language, to refer to the *sides* of a cube; thus, "a cube has 6 sides." We strongly advise you to avoid this very misleading use of the word *side* and to use instead the mathematical word *face*; thus "a cube has 6 faces." We use the word *side* exclusively to refer to a line segment forming part of the boundary of a face. Thus, for us, a cube has 6 faces and each face has 4 sides. The 24 sides are joined in pairs to form the 12 edges of the cube.

Table 5.1.

A polyhedron with:	is called a:
4 faces	tetrahedron
5 faces	pentahedron
6 faces	hexahedron
7 faces	heptahedron
8 faces	octahedron
9 faces	nonahedron
10 faces	decahedron
12 faces	dodecahedron
14 faces	tetracaidecahedron (*cai* means "and")
15 faces	pentacaidecahedron
16 faces	hexacaidecahedron
20 faces	icosahedron

better-known polyhedra. You will note that the part preceding *hedron* designates the number of faces (*poly* means "many"). In the case of a convex polyhedron, each of these faces can be used as a base when a model is set on a table. This explains the use of the generic term *hedron*, which is the Greek word meaning "base" or "seat."

We define a ***regular convex polygon*** (with 3 or more sides) to be a polygon with all sides equal and all angles equal (you may recall, from elementary geometry, that only in the case of the triangle does each of these conditions imply the other). So now we ask: ***What would be an appropriate analogous requirement for a regular polyhedron?***

We reason that, since the faces of polyhedra are analogous to the sides of polygons, it should make sense to require that every face on a regular polyhedron should be the same regular polygon. We can readily agree, however, that this wouldn't be a strong enough requirement. For example, observe that the polyhedra that were constructed entirely of equilateral triangles, namely, the convex deltahedron, are not all regular. Thus, if you view the triangular dipyramid from the top vertex you see 3 triangles meeting, but if you view it from a side vertex you see 4 triangles meeting. A regular polyhedron should surely look the same when viewed from any vertex (or from any edge or from any face). So it would be reasonable to require that the arrangement of the polygonal faces on a regular polyhedron should be exactly the same at every vertex. Let us impose this extra restriction and then see, first of all, if any polyhedra exist that satisfy our requirements and, second, if the resulting polyhedra deserve to be called "regular."

Following this idea we let p stand for the number of sides of each regular face. Then we know, from Section 3.1, that the interior angle of the regular p-gon,

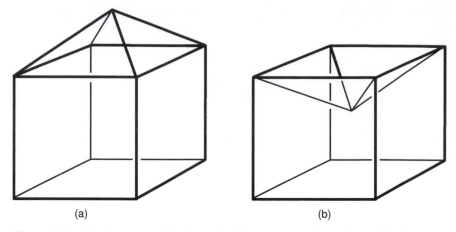

Figure 5.6   (a) Convex polyhedron. (b) Non-convex polyhedron with the same face angles at its vertices as its convex cousin at the left.

measured in radians, is

$$\frac{(p-2)\pi}{p}.$$

Now if we consider the arrangement of q of these regular p-gons about a single vertex, we see that the sum of the face angles about that vertex is

$$q\left(\frac{(p-2)\pi}{p}\right).$$

Notice that $p \geq 3$. However, we also have $q \geq 3$, because at least 3 faces must come together at each vertex.

The concept of convexity now becomes important. It is a fact, first remarked by Euclid (which we illustrate but do not prove), that the sum of the face angles about any vertex on a *convex* polyhedron must be less than 2π. All the models you have constructed so far in this chapter illustrate this fact. However, we see from Figure 5.6(b) that the converse is not true, that is, the sum of the face angles at every vertex may be less than 2π without the polyhedron being convex.

So now let us look for values of p and q, with $p \geq 3$ and $q \geq 3$, such that

$$q\left(\frac{(p-2)\pi}{p}\right) \leq 2\pi$$

or, after dividing by π and multiplying by p, such that

$$pq - 2q < 2p.$$

Straightforward algebra then gives the following sequence of equivalent inequalities:

$$pq - 2p - 2q < 0$$
$$pq - 2p - 2q + 4 < 4$$
$$p(q - 2) - 2(q - 2) < 4$$
$$(p - 2)(q - 2) < 4.$$

Notice that this last expression is symmetric in p and q – and thus if you find any values for p and q that satisfy this inequality, you can find the "complementary" solution by exchanging the values for p and q. Notice also that since $p \geq 3$ and $q \geq 3$, it follows that $p - 2 \geq 1$ and $q - 2 \geq 1$. Thus if $(p - 2)(q - 2) < 4$ then $(p - 2)(q - 2)$ must be 1, 2, or 3. When the product $(p - 2)(q - 2)$ is 1 we must have

$$p - 2 = 1 \quad \text{and} \quad q - 2 = 1$$

so that we obtain

$$p = 3 \quad \text{and} \quad q = 3.$$

If $(p - 2)(q - 2) = 2$, then either

$$p - 2 = 2 \quad \text{and} \quad q - 2 = 1$$

or

$$p - 2 = 1 \quad \text{and} \quad q - 2 = 2.$$

Thus, in the first instance, $p = 4$ and $q = 3$; in the second instance, $p = 3$ and $q = 4$.

Finally, if $(p - 2)(q - 2) = 3$, then either

$$p - 2 = 3 \quad \text{and} \quad q - 2 = 1$$

or

$$p - 2 = 1 \quad \text{and} \quad q - 2 = 3.$$

The first equations give $p = 5$ and $q = 3$; the second equations yield $p = 3$ and $q = 5$.

Thus we see that the only possible solutions to our inequalities

$$(p - 2)(q - 2) < 4, \qquad p, q \geq 3$$

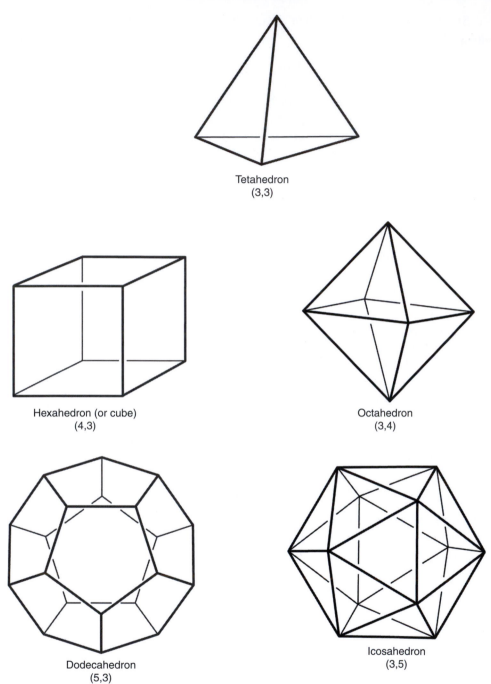

Figure 5.7   The Platonic solids, The notation (p, q) means that each face is a regular p-gon and q faces come together at each vertex.

are

$$p = 3, \qquad q = 3$$
$$p = 3, \qquad q = 4$$
$$p = 4, \qquad q = 3$$
$$p = 3, \qquad q = 5$$
$$p = 5, \qquad q = 3.$$

In the form of a table, these give the following values for the pair (p, q):

p	q
3	3
3	4
4	3
3	5
5	3

In fact, each solution corresponds to an actual polyhedron – and you have already been given instructions for constructing them. Each value of (p, q) determines exactly one polyhedron. The polyhedra in this very special set are known as the **Platonic solids** (see Figure 5.7). We remind you that, as described, they are not really solid because they do not contain their interiors. But we also note, with some pleasure, that they satisfy our criteria for regularity; indeed, nobody could conceivably dispute the claim that they are regular! Thus we adopt our provisional definition of regularity, and we are in the happy position of knowing that these are all of the regular polyhedra – there are just the five Platonic solids.

Instructions for braiding Platonic solids from straight strips appear in Chapter 9.

6

Constructing dipyramids and rotating rings from straight strips of triangles

Required materials

- About 10 yards (10 meters) of 2-inch (5 cm) gummed mailing tape (or a longer length of wider tape if you want larger models). The glue on the tape should be the type that needs to be moistened to become sticky. Don't try to use tape that is sticky to the touch when it is dry – if you do you will find the experience very frustrating.
- Two different colors of brightly colored wrapping paper or butcher paper
- Scissors
- Sponge (or washcloth)
- Shallow bowl
- Water
- Hand towel (or rag)
- Some big (heavy) books
- Some bobby pins, or rubber bands

6.1 Preparing the pattern piece for a pentagonal dipyramid

Begin by folding some of the gummed mailing tape using the $U^1 D^1$-folding procedure as shown in Section 2.3. Do this with the gummed side *up*, so you can see your fold lines better – and so that the paper will "curl" in the right direction when you assemble the model. Continue folding until you have 40 or more triangles. Cut or carefully tear the tape on the last fold line. Then cut off, or tear off, a strip containing 31 triangles counting from this end (that is, *not* from the end from which you began the folding).

Next place your strip of 31 equilateral triangles so that one end appears as shown in Figure 6.1 with the gummed side *down*. Mark the first and eighth triangles *exactly* as shown (note the orientation of each of the letters within their respective triangles).

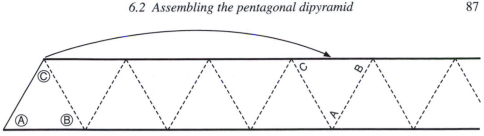

Figure 6.1 Left-hand end of pattern piece.

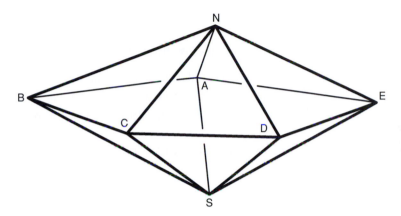

Figure 6.2 Completed pentagonal dipyramid.

6.2 Assembling the pentagonal dipyramid

Begin by placing the first triangle *over* the eighth triangle so that the corner labeled Ⓐ is over the corner labeled A, Ⓑ is over the corner labeled B, and Ⓒ is over the corner labeled C. Hold these two triangles together, in that position, and observe that you have the beginning of a double pyramid for which there will be five triangles above and five triangles below the horizontal plane of symmetry, as shown in Figure 6.2. You may wish to secure the overlapping triangles by moistening just a small part of the center of the gummed triangle.

Now you can hold the model up and let the long strip of triangles fall around this frame. If the strip is folded well, the remaining triangles will fall into place. When you get to the last triangle, there will be a crossing of the strip that the last triangle can tuck into, and *your pentagonal dipyramid is complete*! See Figure 6.2.

If you have trouble because the strip doesn't seem to fall into place, there are two frequent explanations. The first (and more likely) problem is caused by not folding the crease lines firmly enough. This situation is easily remedied by refolding each crease line with more conviction, and it is sometimes helpful to fold each crease line in both directions (so that each fold line is scored as both a *mountain* and a *valley* fold). The second common difficulty occurs when the tape seems too short to reach around the model and tuck in. This happens because we're dealing with the *real world*, where paper has some thickness; we're not doing geometry that is perfect in the mind, where planes have zero thickness. This problem can be remedied by trimming off a tiny amount from each edge of the tape. What this does is to truncate at the vertices the finished model; but since only a small amount is trimmed off, the effect on the appearance of the finished model is not noticeable.

6.3 Refinements for dipyramids

If you want to make a more attractive model, you may glue the strip of triangles onto a piece of *colored* paper. To do this, first prepare the piece of paper onto which you plan to glue the *prefolded* strip. Make certain it is long enough and that it all lies on a flat surface.

Place a sponge (or washcloth) in a bowl. Add water to the bowl so that the top of the sponge is very moist (squishy).[1] Moisten one end of the strip of triangles by pressing it onto the sponge; then, holding that end (yes, it's messy!), pull the rest of the strip across the sponge. Make certain the entire strip gets wet and then place it on the colored paper. Use a hand towel (or rag) to wipe up the excess moisture and to smooth the tape into contact with the colored paper.

Put some books on top of the tape so that it will dry flat. When the tape is dry, cut out the pattern piece, *trimming off a small amount of the gummed tape* (about $\frac{1}{16}$ to $\frac{1}{8}$ of an inch – 1 to 2 mm) from the edge as you do so (this serves to make the model look neater and, more importantly, allows for the increased thickness produced by gluing the strip to another piece of paper). Refold the piece so that the mountain folds (producing raised ridges) are on the colored side of the paper.

You can now construct the polyhedron exactly as before – except that now you probably won't need to label any of the triangles. If you find it difficult to make the last triangle tuck in because it won't reach, you should trim off a little more from each edge.

Next, observe from Figure 6.2 that the completed pentagonal dipyramid has a well-defined equatorial pentagon ABCDE going around the middle. Each edge of this equatorial pentagon is incident with exactly two triangular faces (for instance, AB is incident with the face NAB and face SAB); and each triangular face is incident with exactly one edge of the equatorial pentagon. The two faces sharing

---

[1] Perhaps we should have told you to wear some very old clothes when we mentioned the optional materials!

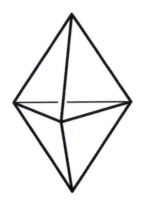

Figure 6.3   Completed triangular dipyramid.

an equatorial edge are said to be **associated**. If the faces are labeled at random with the digits 0, 1, 2, 3, 4, 5, 6, 7, 8, 9, then we may use the pentagonal dipyramid as a die. When we throw the pyramid, it comes to rest on one face, and we declare the number on the *associated* face to be the result of the throw. In this way this die becomes a device for generating random numbers in base 10. There is a slight bias due to the fact that all but one of the faces of the die are covered by exactly three triangles, from our original strip; whereas one face is covered by four triangles. However, this bias can be virtually eliminated by trimming off half of the first and last triangles (so that the pattern piece becomes a rectangle) and pretending the whole triangle is there when you assemble it.

You may wish to figure out how to make the analogous construction of a *triangular dipyramid* from a single strip of 19 equilateral triangles. If you want a fair random-number generator for the numbers 0, 1, 2, 3, 4, 5, trim off half of the first and last triangles. Knowing that the finished model should appear as shown in Figure 6.3, and that you should begin by forming the *top* three faces with one end of the strip, should get you off to a good start.

You may discover that there are ways of constructing these dipyramids with fewer than the number of triangles specified, but the real question is – **Will they be balanced, in the sense that every face is covered by the same number of triangles?** It is not a difficult question – and the answer is therefore left to the reader to discover.

It is an interesting fact that of the 8 convex deltahedra discussed in Section 5.2 only the triangular dipyramid and the pentagonal dipyramid can be constructed with a single strip. To see why this is true take any convex deltahedron, other than one of the dipyramids, and wrap a strip of triangles around it – you will readily see that for any one of these models, there is no way to wrap the strip about its surface so that every edge is covered by the strip.

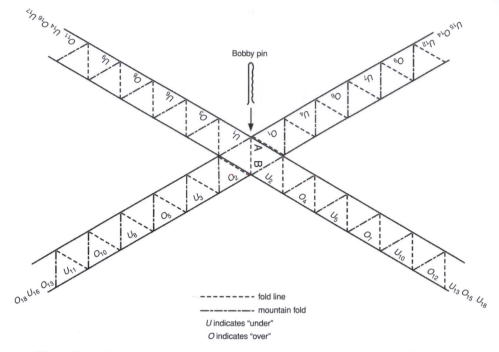

Figure 6.4   The strips for the rotating ring of tetrahedra ready for braiding. (Note that the letters O and U alternate, and the differences between the successive subscripts form a sequence of period 4, namely $\cdots 1232 \cdots$.)

6.4 Constructing braided rotating rings of tetrahedra

Prepare 2 strips, of at least 50 equilateral triangles each, by folding U^1D^1, as illustrated in Section 2.3. Then glue these folded strips to colored paper and cut them out (see Section 6.3 for practical hints about precisely how to do this to get the best results).

Before beginning the construction of the rotating ring, take each pattern piece and fold the paper, very firmly in both directions, so that the completed model will flex more easily. You simply cannot overdo this step – in our experience a rotating ring of tetrahedra improves with age due to the increased flexibility of the edges.

The construction goes as follows:

1. On some fold line near the middle of each strip, cross the 2 strips over each other and secure them with a bobby pin to form a sideways "X" as shown in Figure 6.4.
2. Label the strips *exactly* as shown.
3. To construct the first tetrahedron, lift the edge AB with the bobby pin and slide triangle U_1 underneath O_1; next slide triangle U_2 underneath triangle O_2. Notice that the edges marked — - — - — will always be edges on the tetrahedron that are not attached to any other tetrahedron.

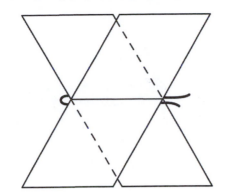

Figure 6.5   The nearly finished rotating ring.

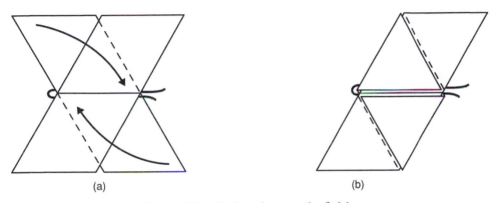

(a)   (b)

Figure 6.6   Getting closer to the finish.

4. Move the bobby pin to the opposite edge of the completed tetrahedron so the strips won't slip apart.
5. Separate the four ends so that U_3, O_3 are on one side and U_4, O_4 are on the other side, forming a sideways "X" under the tetrahedron like the initial configuration in Figure 6.4.
6. Now repeat the braiding process for the triangles marked U_3, O_3 and U_4, O_4.
7. Again, move the bobby pin to the edge of the new tetrahedron, separate the strips, and repeat the process for the triangles U_5, O_5, and U_6, O_6, and so on.
8. When you have braided 10 tetrahedra, trim off all but two triangles from each of the four loose ends at the last edge (which should be secured with the bobby pin).
9. The arrangement of the strips at the end should look like Figure 6.5.
10. Fold back the two triangles, as shown in Figure 6.6(a), to get the arrangement shown in Figure 6.6(b).
11. Place the edge shown as A'B' next to the edge AB on the first tetrahedron and tuck the two triangles into the openings (or slots) on the edge of this first tetrahedron so that they go in the *backwards* direction. Figure 6.7 shows how the last triangle should look as it slides into the slot. Figure 6.8 shows the completed rotating ring of tetrahedra.

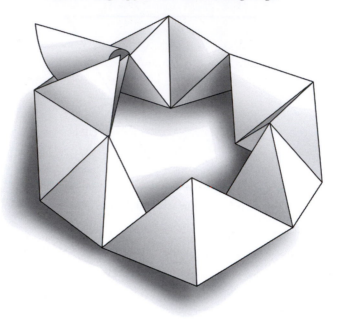

Figure 6.7   Almost there.

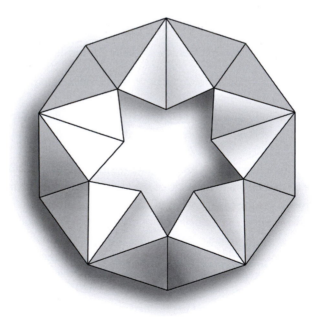

Figure 6.8   The completed rotating ring of 10 tetrahedra.

Practical hints

A common mistake is to fail to cross the strips properly while braiding the model. The result is that, although the model looks fine, it comes apart as you rotate it. If this happens, undo it and rebraid it. A good rule to remember is that when two strips meet at a crossing

> the strip that came from underneath goes over next

and

> the strip that was on top goes underneath next.

When you are flexing a new rotating ring of tetrahedra, the paper may have a tendency to buckle. You must be gentle with it when pushing the offending edge back into its proper position. After a while two things will happen. First, you will become more adept at turning the model (because you will have gained a better understanding of its mechanical properties) and, second, the model itself will become more pliable.

6.5 Variations for rotating rings

We were not obliged to use exactly 10 tetrahedra. In fact, a rotating ring can be constructed with just 8 regular tetrahedra (but not 6 – try it and you'll see why). We chose to use 10 tetrahedra because we like the symmetry of the final model.

Rings containing 22 or more tetrahedra can be tied in various knots before joining the first and last tetrahedra to each other. These rotating knots form interesting configurations whose twisting motions are almost hypnotic.

Why do you suppose it was possible to make this model?

Observe that the regular tetrahedron is the only one of the Platonic solids whose opposite edges, when extended, do *not* form parallel lines. In fact, it is this property that makes it possible to use these tetrahedra to construct a ring that will rotate. You can produce non-rigid rings by joining opposite edges of other kinds of polyhedra (cubes, for example), but they will not rotate unless the opposite edges used for joining the polyhedra lie on non-parallel lines. Moreover, if the pair of joining edges on each of several similar polyhedra are not on lines that are at right angles to each other in space, the symmetries of the final rotating ring may be quite unusual. We suggest that you experiment with different numbers of tetrahedra in your rotating rings. Try tying knots before joining the first and last tetrahedra and, if you are really interested, make some rotating rings with other polyhedra. Figure 6.9, which resembles a holiday wreath in appearance, is made from 14 hexacaidecadeltahedra.

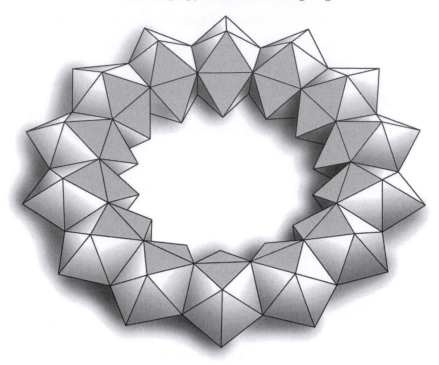

Figure 6.9   A rotating ring of 14 hexacaidecadeltahedra.

6.6 More fun with rotating rings

Suppose you wish to number the visible faces of your rotating ring of 10 tetrahedra with the consecutive numbers $1, 2, 3, \ldots, 40$. Now

$$\sum_{i=1}^{40} i = 1 + 2 + 3 + \cdots + 40 = 820.$$

A very nice way of seeing this is to rearrange the sum with the first half of the numbers ascending and the second half descending, placed directly underneath it, and then add vertically. The arrangement then appears as

$$
\begin{array}{c}
1 + 2 + 3 + \cdots + 19 + 20 \\
40 + 39 + 38 + \cdots + 22 + 21 \\
\hline
41 + 41 + 41 + \cdots + 41 + 41
\end{array}
$$

We readily see that 41 occurs precisely 20 times in the third line. Thus

$$1 + 2 + 3 + \cdots + 40 = 20(41) = 820.$$

Now, since 820 is exactly divisible by 10 (the number of tetrahedra in our ring), it is sensible to ask whether or not it is possible to number the faces of the rotating ring with the numbers from 1 to 40 so that the numbers on the four faces of each tetrahedron will sum to 82. We believe the reader will be able to answer this question – and that you will also be able to ask and answer the corresponding question for rings involving other numbers of tetrahedra. Rouse Ball reported in *Mathematical Recreations and Essays* [2] that a certain R. V. Heath, in answering the corresponding question of a rotating ring of 8 tetrahedra, showed how one could assign the numbers 1, 2, 3, . . . , 32 (which sum to 528) to the 32 faces so that the 4 faces of each tetrahedron sum to 66, and "corresponding" faces, one from each tetrahedron, sum to 132.

Suppose you consider a rotating ring of N tetrahedra. The sum of the integers 1, 2, 3, . . . , $4N$ is $2N(4N + 1)$. Our original question has to do with assigning one of these integers to each face so that the sum for each tetrahedron is $2(4N + 1)$. Heath's refinement would require that the sum of "corresponding" faces be $\frac{N(4N+1)}{2}$; plainly to achieve this refinement, N must be even.

Doris Schattschneider and Wallace Walker have produced a monograph [74] that includes die-cut nets from which fascinating solids and rotating rings can be constructed that have Escher-type designs on their faces. Their kit would surely give you many ideas for decorating not only rotating rings but also many other polyhedra.

Some history about rotating rings

One of the authors, JP, first encountered these delightful models in Ball's book, *Mathematical Recreations and Essays* [2], where it was reported that these models were originally discovered independently by J. M. Andreas and R. M. Stalker. Ball also observed that the rotating rings have $2n$ vertices, $6n$ edges (of which $2n$ coincide in pairs) and $4n$ triangular faces for n equal to 6, 8, 9, 10, 11, . . . What is new in our approach is the braiding of these models from two straight strips (see [58]), rather than building the individual tetrahedra from nets and then attaching them to each other with some kind of flexible material. We first wrote about this technique as a team in [33].

7

Continuing the paper-folding and number-theory threads

7.1 Constructing an 11-gon

Suppose we want to construct a regular star $\left\{\frac{11}{3}\right\}$-gon. Then, of course, $b = 11$, $a = 3$ in the notation used in Section 2.5, and we proceed as we did when we wished to construct the regular convex 7-gon in that chapter – we adopt our **optimistic strategy** (which means that we *assume* we've got what we want and, as we will show, we then actually *get* an arbitrarily good approximation to what we want!). Thus we assume we can fold the desired putative angle of $\frac{3\pi}{11}$ at A_0 (see Figure 7.1(a)) and we adhere to the same principles that we used in constructing the regular 7-gon, that is, we adopt the following rules:

(1) Each new crease line goes in the forward (left to right) direction along the strip of paper.
(2) Each new crease line always *bisects* the angle between the last crease line and the edge of the tape from which it emanates.
(3) The bisection of angles at any vertex continues until a crease line produces an angle of the form $\frac{a'\pi}{b}$ where a' is an *odd* number; then the folding stops at that vertex and commences at the intersection point of the last crease line with the other edge of the tape.

Once again the **optimistic strategy** works; and following this procedure results in tape whose angles converge to those shown in Figure 7.1(b). We could denote this folding procedure by $D^1U^3D^1U^1D^3U^1$, interpreted in the obvious way on the tape – that is, the first exponent "1" refers to the 1 bisection (producing a line in a downward direction) at the vertices A_{6n} (for $n = 0, 1, 2, \ldots$) on the top of the tape; similarly, the "3" refers to the 3 bisections (producing creases in an upward direction) made at the bottom of the tape through the vertices A_{6n+1}; etc. However, since the folding procedure is *duplicated* half-way through, we can abbreviate the notation and write simply $\{1, 3, 1\}$, with the understanding that we alternately fold

96

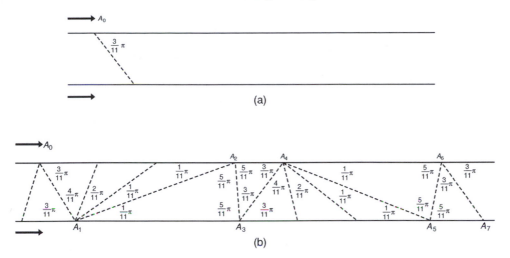

Figure 7.1   Folding tape to produce an $\{\frac{11}{3}\}$-gon.

from the top and bottom of the tape as described, with the *number* of bisections at each vertex running, in order, through the values $1, 3, 1, \ldots$ We call this a ***primary folding procedure of period 3 or a 3-period folding***.

A proof of convergence for the general folding procedure of arbitrary period may be given that is similar to the one we gave for the primary folding procedure of period 2. Or one could revert to an error-correction type of proof like that given for the 7-gon in Section 3.3. We leave the details to the reader, and explore here what we can do with this $(1, 3, 1)$-tape. First, note that, starting with the putative angle $\frac{3\pi}{11}$ at the top of the tape, we produce a putative angle of $\frac{\pi}{11}$ at the bottom of the tape, then a putative angle of $\frac{5\pi}{11}$ at the top of the tape, then a putative angle of $\frac{3\pi}{11}$ at the *bottom* of the tape, and so on. Hence we see that we could use this tape to fold a star $\{\frac{11}{3}\}$-gon, a convex 11-gon, and a star $\{\frac{11}{5}\}$-gon. More still is true; for, as we see, if there are crease lines enabling us to fold a star $\{\frac{11}{a}\}$-gon, there will be crease lines enabling us to fold star $\{\frac{11}{2^k a}\}$-gons, where $k \geq 0$ takes any value such that $2^{k+1}a < 11$. These features, described for $b = 11$, would be found with any odd number b. However, this tape has a special symmetry as a consequence of its *odd* period; namely, if it is "flipped" about the horizontal line half-way between its parallel edges, the result is a *translate* of the original tape. As a practical matter this special symmetry of the tape means that we can use either the top edge or the bottom edge of the tape to construct our polygons. On tapes with an *even* period the top edge and the bottom edge of the tape are not translates of each other (under the horizontal flip), which simply means that care must be taken in choosing the edge of the tape to be used to construct a specific polygon. Figures 7.2(a), (b) show the completed $\{\frac{11}{3}\}$-, $\{\frac{11}{4}\}$-gons, respectively.

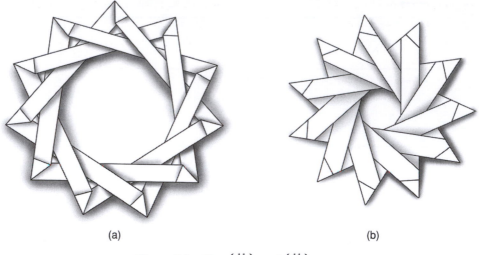

Figure 7.2 Star $\left\{\frac{11}{3}\right\}$- and $\left\{\frac{11}{4}\right\}$-gons.

Now, to set the scene for the number theory of Section 7.2, let us look at the patterns in the *arithmetic* of the computations when $a = 3$ and $b = 11$. Referring to Figure 7.1(b) we observe that

The smallest angle to the right of A_n where	is of the form $\frac{a}{11}\pi$ where	and the number of bisections at the *next* vertex[1]
$n = 0$	$a = 3$	$= 3$
1	1	1
2	5	1
3	3	3
4	1	1
5	5	1

We could write this in shorthand form as follows:

$$(b =)11 \begin{vmatrix} (a =)3 & 1 & 5 \\ 3 & 1 & 1 \end{vmatrix}. \tag{7.1}$$

Observe that, had we started with the putative angle of $\frac{\pi}{11}$, then the *symbol* (7.1) would have taken the form

$$(b =)11 \begin{vmatrix} (a =)1 & 5 & 3 \\ 1 & 1 & 3 \end{vmatrix}. \tag{7.1'}$$

---

[1] Notice that, referring to Figure 7.1(b), to obtain an angle of $\frac{3\pi}{11}$ at $A_0, A_6, A_{12}, \ldots$, the folding instructions would more precisely be $U^3 D^1 U^1 D^3 U^1 D^1 \cdots$. But we don't have to worry about this distinction here.

In fact, it should be clear that we can *start anywhere* (with $A = 1, 3$ or 5) and the resulting symbol, analogous to (7.1′), will be obtained by cyclic permutation of the matrix component of the symbol, placing our choice of a in the first position along the top row.

In general, suppose we wish to fold a $\{\frac{b}{a}\}$-gon, with b, a odd and $a < \frac{b}{2}$. Then we may construct a symbol[2] as follows. Let us write

$$b \begin{vmatrix} a_1 & a_2 & \cdot & \cdot & \cdot & a_r \\ k_1 & k_2 & \cdot & \cdot & \cdot & k_r \end{vmatrix} \tag{7.2}$$

where $b, a_i (a_1 = a)$ are odd, $a_i < \frac{b}{2}$, and

$$b - a_i = 2^{k_i} a_{i+1}, \qquad i = 1, 2, \ldots, r, \qquad a_{r+1} = a_1. \tag{7.3}$$

We emphasize, as we will prove at the beginning of Section 7.2, that, given any two odd numbers a and b, with $a < \frac{b}{2}$, there is always a completely determined unique symbol (7.2) with $a_1' = a$. At this stage, we do not assume that $\gcd(b, a) = 1$, but we have assumed that the list $a_1, a_2, \ldots, a_r$ is without repeats. Indeed, if $\gcd(b, a) = 1$ we say that the symbol (7.2) is **reduced**, and if there are no repeats among the a_i's we say that the symbol (7.2) is **contracted**. (It is, of course, theoretically possible to consider symbols (7.2) in which repetitions among the a_i are allowed.) We regard (7.2) as encoding the general folding procedure to which we have referred.

Example 7.1 If we wish to fold a 17-gon we may start with $b = 17, a = 1$ and construct the symbol

$$(b =)17 \begin{vmatrix} (a =)1 \\ 4 \end{vmatrix},$$

which tells us that folding $U^4 D^4$ will produce tape (usually denoted (4, 4)-tape) that can be used to construct a FAT 17-gon. In fact, this tape can also be used to construct FAT

$$\left\{\tfrac{17}{2}\right\}\text{-,} \quad \left\{\tfrac{17}{4}\right\}\text{-,} \quad \text{and} \quad \left\{\tfrac{17}{8}\right\}\text{-gons.}$$

However, if we wish to fold a $\left\{\tfrac{17}{3}\right\}$-gon we start with $b = 17, a = 3$ and construct the symbol

$$(b =)17 \begin{vmatrix} (a =)3 & 7 & 5 \\ 1 & 1 & 2 \end{vmatrix}$$

---

[2] More exactly, a 2-symbol. Later on, we will introduce the idea of a more general t-symbol, $t \geq 2$.

which tells us to fold $U^1 D^1 U^2 D^1 U^1 D^2$ – or, more simply, to use the $(1, 1, 2)$-folding procedure – to produce $(1, 1, 2)$-tape from which we can fold the FAT $\left\{\frac{17}{3}\right\}$-gon. Again, we get more than we initially sought, since we can also use the $(1, 1, 2)$-tape to construct

$$\text{FAT} \quad \left\{\tfrac{17}{6}\right\}\text{-,} \quad \left\{\tfrac{17}{7}\right\}\text{-,} \quad \text{and} \quad \left\{\tfrac{17}{5}\right\}\text{-gons.}$$

We can combine all the possible symbols for $b = 17$ into one **complete** symbol, adopting the notation

$$17 \left|\begin{array}{c|ccc} 1 & 3 & 7 & 5 \\ 4 & 1 & 1 & 2 \end{array}\right|.$$

Check your understanding of the complete symbol for $b = 93$ by doing the calculations to fill in the blanks below:

$$93 \left|\begin{array}{cccc|cccc|cccccc} 1 & 23 & 35 & 29 & 5 & 11 & 41 & 13 & 7 & 43 & 25 & 17 & 19 & 37 \\ 2 & 1 & 1 & 6 & 3 & ? & ? & 4 & 1 & ? & ? & ? & ? & ? \end{array}\right|.$$

(Notice that no multiples of 3 or 31 appear in the top row. Why not?)

You may wish to calculate other complete symbols and study them, along with the complete symbols for 17 and 93 above, to see if you notice any consistent patterns or features in them. We will discuss these features in more detail in Chapter 17.

Suppose we were to allow unreduced symbols. For example, let us calculate the symbol with $b = 51$, $a = 9$. We would get

$$51 \left|\begin{array}{ccc} 9 & 21 & 15 \\ 1 & 1 & 2 \end{array}\right|.$$

What do you notice? You should now see why it is pointless to allow unreduced symbols.

*7.2 The quasi-order theorem

The first claim we have to substantiate, made in the previous section, is that, given positive odd integers b, a with $a < \frac{b}{2}$, there is always a unique contracted symbol

$$b \left|\begin{array}{ccccc} a_1 & a_2 & \cdot & \cdot & \cdot & a_r \\ k_1 & k_2 & \cdot & \cdot & \cdot & k_r \end{array}\right|, \qquad a_1 = a, \qquad a_i \neq a_j \qquad \text{if} \quad i \neq j, \quad (7.4)$$

where each a_i is odd, $a_i < \frac{b}{2}$, and, as before,

$$b - a_i = 2^{k_i} a_{i+1}, \qquad i = 1, 2, \ldots, r, \quad \text{and} \quad a_{r+1} = a_1. \tag{7.5}$$

We argue as follows. We fix b and let S be the set of positive odd integers $a < \frac{b}{2}$. Given $a \in S$, define a' by the rule

$$b - a = 2^k a', \quad k \text{ maximal.} \tag{7.6}$$

Notice that $k \geq 1$, since $b - a$ is certainly even. We claim that $a' \in S$. First, a' is obviously odd. Second, $2a' \leq 2^k a' = b - a < b$, so $a' < \frac{b}{2}$. Thus (7.6) describes a function $\psi \colon S \longrightarrow S$, such that $\psi(a) = a'$. We will show that ψ is a ***permutation*** of the finite set S; to show this it is certainly enough to exhibit a function $\varphi \colon S \longrightarrow S$ such that $\psi\varphi(a') = a'$. We define φ as follows: given $a' \in S$, let k be *minimal* such that $2^k a' > \frac{b}{2}$ and set $\varphi(a') = a$, where

$$a = b - 2^k a'. \tag{7.7}$$

Notice that $k \geq 1$, since $a' < \frac{b}{2}$, so that a is odd; that $a < \frac{b}{2}$ since $2^k a' > \frac{b}{2}$; and that $b > 2^k a'$, since $2^{k-1} a' < \frac{b}{2}$, so that a is positive. Thus $a \in S$; and comparison of (7.6), (7.7) shows that, as claimed, $\psi\varphi(a') = a'$. Thus ψ is a permutation with inverse permutation φ.

The permutation ψ has one more important property. We write $\psi(a) = a'$, as above, and claim that

$$\gcd(b, a) = \gcd(b, a'). \tag{7.8}$$

For it is clear from (7.6) that if $d \mid b$ and $d \mid a'$ then $d \mid b$ and $d \mid a$. Conversely, if $d \mid b$ and $d \mid a$, d is *odd* and $d \mid 2^k a'$ so $d \mid b$ and $d \mid a'$. Thus if a_1 in (7.4) is coprime to b, so are $a_2, a_3, \ldots, a_r$, and we may, if we wish, confine ψ and φ to the subset S_0 of S consisting of those $a \in S$ which are coprime to b; that is, we may confine ourselves to reduced symbols.

We now use the fact that, given *any* permutation ψ of a finite set S_0 and any $a \in S_0$, then a must generate a ***cycle***, in the sense that, if we iterate ψ, getting

$$a, \psi(a), \psi^2(a), \psi^3(a), \ldots$$

(here $\psi^2(a) = \psi(\psi(a))$, etc. and we may write $\psi^0(a)$ for a) we must eventually repeat, that is, we will find $m > 0$ such that

$$a, \psi(a), \ldots, \psi^{m-1}(a)$$

are all different but $\psi^m(a) = a$. In case this is not clear to you, we give the easy proof. Certainly, since S_0 is finite, we must eventually repeat in the weaker sense that we find $s \geq 0, m > 0$ such that $\psi^s(a) = \psi^{s+m}(a)$. Suppose this is the *first* time we get a repeat. We claim that $s = 0$; for, if not, we have

$$\psi(\psi^{s-1}(a)) = \psi(\psi^{s-1+m}(a)).$$

But ψ is one-one, so $\psi^{s-1}(a) = \psi^{s-1+m}(a)$ and the given repeat wasn't our first repeat. Thus $s = 0$ and $a = \psi^m(a)$. Of course, with regard to (7.4), $m = r + 1$. This completes the proof of our claim.

So we have a universal algorithm for folding a $\left\{ \frac{b}{a} \right\}$-gon, where a, b are coprime odd integers with $a < \frac{b}{2}$. But, from the number-theoretic point of view, it turns out that we have much more. For, reverting to (7.4), let

$$k = \sum_{i=1}^{r} k_i \quad \text{(we may call this the } \textit{fold-total}\text{)}.$$

Now we can prove our theorem.

The quasi-order theorem[3] *Suppose that (7.4) is not only contracted but also reduced. Then the quasi-order of $2 \bmod b$ is k. That is, k is the smallest positive integer such that*

$$2^k \equiv \pm 1 \bmod b.$$

In fact, $2^k \equiv (-1)^r \bmod b$.

Proof. The proof is really a triumph of technique! First, we find it convenient to think in terms of the φ-function rather than the ψ-function. Thus we work **backwards** in constructing our symbol (7.4). Also we will find it convenient to repeat the initial number a_1. Precisely, we write our **modified symbol** as

$$b \begin{pmatrix} c_1 & c_2 & c_3 & \cdot & \cdot & \cdot & c_r & c_1 \\ & \ell_1 & \ell_2 & \ell_3 & \cdot & \cdot & \cdot & \ell_r & \end{pmatrix} \qquad (7.9)$$

where[4] (compare (7.5) and (7.7)) ℓ_i is minimal such that $2^{\ell_i} c_i > \frac{b}{2}$ and

$$b - c_{i+1} = 2^{\ell_i} c_i, \qquad i = 1, 2, \ldots, \qquad r(c_{r+1} = c_1). \qquad (7.10)$$

We set $\ell = \sum_{i=1}^{r} \ell_i$, and our first task will be to prove that

$$2^{\ell} \equiv (-1)^r \bmod b. \qquad (7.11)$$

---

[3] You may like to know that this result is fairly new; we first published it in 1983 (see [27, 30]).
[4] In fact, if we compare (7.4) and (7.9), $k_i = \ell_{r+1-i}$, $a_i = c_{r+2-i}$.

To this end, consider the $(\ell + 1)$ numbers, all less that $\frac{b}{2}$,

$$c_1, 2c_1, \ldots, 2^{\ell_1-1}c_1, c_2, 2c_2, \ldots, 2^{\ell_2-1}c_2, c_3, \ldots, c_r, \ldots, 2^{\ell_r-1}c_r, c_1. \qquad (7.12)$$

In this sequence there are r places where we *switch* from c_i to c_{i+1}. If we rewrite the sequence (7.12) as

$$n_1, n_2, n_3, \ldots, n_{\ell+1},$$

then

$$n_{j+1} = \begin{cases} 2n_j & \text{if there is no switch} \\ -2n_j \bmod b & \text{at a switch, by (7.10).} \end{cases} \qquad (7.13)$$

Since there are r switches, we conclude from (7.13) that

$$n_{\ell+1} \equiv (-1)^r 2^\ell n_1 \bmod b. \qquad (7.14)$$

But $n_{\ell+1} = n_1 = c_1$, and c_1 is coprime to b. Thus (7.14) implies that

$$2^\ell \equiv (-1)^r \bmod b,$$

which is (7.11).

To show that ℓ is the quasi-order of $2 \bmod b$, we must show that, for every positive $m < \ell$, the congruence

$$2^m \equiv \pm 1 \bmod b \qquad (7.15)$$

is *false*. Now (7.15) implies, in the light of (7.13), that $n_{m+1} \equiv \pm c_1 \bmod b$, with $m + 1 < \ell + 1$. We first show that

$$n_{m+1} \equiv c_1 \bmod b$$

is impossible. Now, as we have remarked, it follows from the definition of the φ-function that all the numbers n_j in the sequence (7.12) satisfy $n_j < \frac{b}{2}$. Thus

$$n_{m+1} \equiv c_1 \bmod b \quad \text{implies} \quad n_{m+1} = c_1.$$

But either n_{m+1} is even or it is some c_i different from c_1. Thus, since the symbol (7.9) is contracted and, of course, c_1 is odd, $n_{m+1} = c_1$ is impossible.

Finally we show that $n_{m+1} \equiv -c_1 \bmod b$ is impossible. For $n_{m+1} + c_1$ is a positive integer less than b, so it is not divisible by b. This completes the proof of the theorem. $\qquad \square$

The quasi-order theorem is striking in that, given b, we compute k starting with *any* a which is odd, less than $\frac{b}{2}$, and coprime to b. Of course, the choice $a = 1$

is always available and is the one to make when we seek folding instructions for producing a regular convex b-gon.

*7.3 The quasi-order theorem when $t = 3$

In Section 4.1 we emphasized the importance of generalization and, almost throughout, replaced the base 2 by an arbitrary base $t \geq 2$. Such a generalization is perfectly possible here and leads to a generalization of the quasi-order theorem. However, the algorithm for executing the ψ-function is more complicated, so we will not go into details here. For the time being we will just show you what happens if $t = 3$. We form our 3-symbol starting with a and b coprime to 3, a coprime to b, and $a < \frac{b}{a}$ (notice this: we do *not* require $a < \frac{b}{3}$). The (modified) ψ-function now allows us to consider $b - a$ or $b + a$. Exactly one of these is divisible by 3, and that is the one we take. We adjoin a third row to our symbol; in this row we write 1 if we took $b - a$ and 0 if we took $b + a$. The second row records the number of times we took the factor 3 out of $b - a$ or $b + a$ to get a'. Here's an example. We have (with $t = 3$, $b = 19$ and $a_1 = 1$) the 3-symbol

$$19 \begin{vmatrix} 1 & 2 & 7 & 4 & 5 & 8 \\ 2 & 1 & 1 & 1 & 1 & 3 \\ 1 & 0 & 1 & 1 & 0 & 0 \end{vmatrix}.$$

This symbol records the following calculations:

$$19 - 1 = 3^2 \cdot 2,$$
$$19 + 2 = 3^1 \cdot 7,$$
$$19 - 7 = 3^1 \cdot 4,$$
$$19 - 4 = 3^1 \cdot 5,$$
$$19 + 5 = 3^1 \cdot 8,$$
$$19 + 8 = 3^3 \cdot 1.$$

We will be content, here, simply to state the general quasi-order theorem in the case $t = 3$. (See Chapter 17 for details in base t.) We select positive integers b, a such that $a < \frac{b}{2}$, b is prime to 3, and $3 \nmid a$. We may then form a 3-symbol

$$b \begin{vmatrix} a_1 & a_2 & \cdot & \cdot & \cdot & a_r \\ k_1 & k_2 & \cdot & \cdot & \cdot & k_r \\ \epsilon_1 & \epsilon_2 & \cdot & \cdot & \cdot & \epsilon_r \end{vmatrix}, \quad \epsilon_i = 0 \text{ or } 1.$$

Let $k = \sum_{i=1}^{r} k_i$, $\epsilon = \sum_{i=1}^{r} \epsilon_i$. Then the quasi-order of $3 \bmod b$ is k and, in fact, $3^k \equiv (-1)^\epsilon \bmod b$.

Reverting to our example above, the conclusion is that the quasi-order of $3 \bmod 19$ is $9 (= 2 + 1 + 1 + 1 + 1 + 3)$ and that, in fact,

$$3^9 \equiv -1 \bmod 19,$$

choosing the minus sign since there is an *odd* number of 1's in the third row.

We see that in constructing a 3-symbol, exactly one of $b - a, b, b + a$ would be divisible by 3; but b isn't divisible by 3. Note that we don't need a third row in the symbol with $t = 2$, because with $t = 2$ we always subtract.

7.4 Paper-folding connections with various famous number sequences

It might impress the reader to know that there are interesting connections between the mathematics of paper-folding and number theory in other parts of mathematics. For example, in [36] we show some surprising connections (involving divisibility criteria) between folding numbers and generalized Fibonacci and Lucas numbers. In fact, the authors of [15], [53], and [66], apply our paper-folding construction methods to algebraic problems, dynamical systems, and diverse geometric applications, respectively.

We will be content in the remainder of this section to just give examples involving Mersenne and Fermat numbers – referring the reader to papers that carry each of the stories further.

There are two rather remarkable examples of our symbol (7.4). One is the complete symbol

$$23 \begin{vmatrix} 1 & 11 & 3 & 5 & 9 & 7 \\ 1 & 2 & 2 & 1 & 1 & 4 \end{vmatrix}$$

telling us that

$$2^{11} \equiv 1 \bmod 23, \quad \text{or} \quad 23 \,|\, 2^{11} - 1. \tag{7.16}$$

This is remarkable because $2^{11} - 1$ is a **Mersenne number**, that is, a number of the form $2^p - 1$, where p is prime. Abbé Mersenne (1588–1648) hoped that all these numbers would be prime; but (7.16) shows that this is not so (of course, this was already known long before the invention of the symbol (7.4)). In [43] we examined this result further and, in the process were led to see, and prove, numerical phenomena like the simple one shown here:

$$
\begin{array}{llll}
\textbf{Case 1:} & 23 \times & 89 = & 2047 \\
\textbf{Case 2:} & 23 \times & 889 = & 20\,447 \\
\textbf{Case 3:} & 23 \times & 8889 = & 204\,447 \\
& \bullet \\
& \bullet \\
& \bullet \\
\textbf{Case } \textbf{\textit{n}}\textbf{:} & 23 \times & 8^k 9 = & 204^k 7,
\end{array}
$$

where we write $\ldots n^k \ldots$ in a base 10 numeral to indicate that the digit n is repeated k times, $k \geq 0$.

The second example is even more remarkable; it is the symbol (7.4) with $b = 641$. We then have

$$
641 \begin{array}{|ccccccccc|}
1 & 5 & 159 & 241 & 25 & 77 & 141 & 125 & 129 \\
7 & 2 & 1 & 4 & 3 & 2 & 2 & 2 & 9
\end{array} \tag{7.17}
$$

telling us that

$$
2^{32} \equiv -1 \bmod 641, \quad \text{or} \quad 641 \mid 2^{32} + 1. \tag{7.18}
$$

This is striking because $2^{32} + 1$ is a **Fermat number**; that is, a number of the form

$$
2^{2^n} + 1.
$$

The French mathematician Pierre Fermat (1601–1665) hoped that all those numbers would be prime; but (7.18) shows that this is not so (the factorizability of $2^{32} + 1$ was first noticed by the Swiss mathematician Leonhard Euler (1707–1783)). In [39] we went further, characterizing the nature of some of the coaches in a complete symbol for numbers b that are products of Fermat numbers.

7.5 Finding the complementary factor and reconstructing the symbol

Apparently, our symbol only gives one factor of $2^{11} - 1$ or $2^{32} + 1$, not the other. We will show that this is not so – we get both factors. The context for this part of the story is the following natural question, which was first raised in [29].

> *Given a folding procedure $(k_1, k_2, \ldots, k_r)$, what polygons do we fold with it?*

The equivalent arithmetical (or algebraic) problem is this:

Given $(k_1, k_2, \ldots, k_r)$, find b, a_1, a_2, $\ldots$, a_r, so that

$$b \begin{vmatrix} a_1 & a_2 & \cdot & \cdot & \cdot & a_r \\ k_1 & k_2 & \cdot & \cdot & \cdot & k_r \end{vmatrix} \tag{7.4}$$

is a reduced and contracted 2-symbol.

Of course, for (7.4) to be contracted, it is necessary that $(k_1, k_2, \ldots, k_r)$ should not consist of repetitions of some (shorter) sequence $(k_1, k_2, \ldots, k_s)$. (This condition is also, in fact, sufficient.)

We now show how to find b, a_1, a_2, $\ldots$, a_r in (7.4). In fact, we solve the simultaneous equations (7.5), that is, $b - a_i = 2^{k_i} a_{i+1}$, $i = 1, 2, \ldots, r$, $a_{r+1} = a_1$, for the "unknowns" $a_1, a_2, \ldots, a_r$, obtaining

$$Ba_i = bA_i, \quad i = 1, 2, \ldots, r, \tag{7.19}$$

where

$$B = 2^k - (-1)^r \tag{7.20}$$

and

$$A_i = 2^{k-k_{i-1}} - 2^{k-k_{i-1}-k_{i-2}} + 2^{k-k_{i-1}-k_{i-2}-k_{i-3}} - \cdots + (-1)^r 2^{k_i} - (-1)^r. \tag{7.21}$$

Let us explain the notation of (7.20), (7.21). First remember that $k = \sum_{i=1}^{r} k_i$. Then, in the expression for A_i, we interpret the subscripts on the k's to be "mod r"; thus, for example, if $i = 1$, then $k_0 = k_r$, $k_{-1} = k_{r-1}$, etc. (This makes good sense since the symbol (7.4) would repeat if allowed, so that we may think of it as written on a cylinder.)

Now B and A_i are determined by $(k_1, k_2, \ldots, k_r)$ and then (7.19) tells us that

$$\frac{b}{a_i} = \frac{B}{A_i}.$$

Thus, since b and a_i are to be coprime, we find them by reducing the fraction $\frac{B}{A_i}$. Notice that this works, that is, the b we get is independent of i, because $\gcd(B, A_i)$ is itself independent of i. The argument is just as for (7.8), being based on the fact that B is odd and, as you may prove from (7.20), (7.21),

$$B - A_i = 2^{k_i} A_{i+1}.$$

If we write d for $\gcd(B, A_i)$, then

$$B = db, \qquad A_i = da_i. \tag{7.22}$$

Let us, as an example, show how to apply our algorithm to fill in the reduced symbol

$$b \begin{vmatrix} a_1 & a_2 & a_3 & a_4 \\ 1 & 2 & 2 & 3 \end{vmatrix}. \tag{7.23}$$

From (7.20) we find $B = 2^8 - 1 = 255$; and from (7.21) we find

$$\begin{aligned} A_1 &= 2^5 - 2^3 + 2^1 - 1 = 25 \\ A_2 &= 2^7 - 2^4 + 2^2 - 1 = 115 \\ A_3 &= 2^6 - 2^5 + 2^2 - 1 = 35 \\ A_4 &= 2^6 - 2^4 + 2^3 - 1 = 55. \end{aligned}$$

Now $\gcd(B, A_1) = 5$; hence, dividing by 5, we find, from (7.22),

$$b = 51, \quad a_1 = 5, \quad a_2 = 23, \quad a_3 = 7, \quad a_4 = 11.$$

If you start with $b = 51$, and $a_1 = 5$, and construct the symbol (7.21) from scratch, you should see, with some satisfaction, that the bottom row of the symbol will be just as in (7.23) – and, of course, the a_i's will be as given above.

Let us make one remark before passing on to the complementary factors for our Mersenne number $2^{11} - 1$ and our Fermat number $2^{32} + 1$. As you saw in the example, once we have found b and a_1 it is easier to complete the symbol (7.23) just by applying the ψ-function (this would also give us a check on our calculation). Thus a more efficient algorithm is this:

(i) calculate B, A_1 from (7.20), (7.21);
(ii) calculate $d = \gcd(B, A_1)$;
(iii) determine b, a_1 from $B = db$, $A_1 = da_1$;
(iv) complete the symbol (7.4) using the ψ-function (that is, $\psi(a_i) = a_{i+1}$, where $b - a_i = 2^{k_i} a_{i+1}$, $i = 1, 2, \ldots, r$, and $a_{r+1} = a_1$).

Now consider again the symbol

$$23 \begin{vmatrix} 1 & 11 & 3 & 5 & 9 & 7 \\ 1 & 2 & 2 & 1 & 1 & 4 \end{vmatrix}.$$

Then (7.20) tells us that $B = 2^{11} - (-1)^6 = 2^{11} - 1$, and so (7.19) tells us that

$$2^{11} - 1 = 23A_1.$$

Thus we will complete the factorization of $2^{11} - 1$ by calculating A_1. From (7.19) this yields

$$A_1 = 2^7 - 2^6 + 2^5 - 2^3 + 2^1 - 1 = 89,$$

so that

$$2^{11} - 1 = 23 \times 89. \tag{7.24}$$

Likewise, applying (7.20) and (7.19) to the symbol (7.17) tells us that $B = 2^{32} + 1$ and $2^{32} + 1 = 641 A_1$, where, by (7.21),

$$A_1 = 2^{23} - 2^{21} + 2^{19} - 2^{17} + 2^{14} - 2^{10} + 2^9 - 2^7 + 1 = 6\,700\,417,$$

so that

$$2^{32} + 1 = 641 \times 6\,700\,417. \tag{7.25}$$

It is striking – don't you agree? – that (7.24) and (7.25) have been obtained without ever expressing $2^{11} - 1$ or $2^{32} + 1$ in base 10 – and it is also striking that the fact that 641 is a factor of $2^{32} + 1$ was established without ever using a number bigger than 641.

It has recently been shown that the smallest prime factor of the Fermat number

$$F_6(= 2^{2^6} + 1) \text{ is } 274\,177;$$

and that the smallest prime factor of the Fermat number

$$F_9(= 2^{2^9} + 1) \text{ is } 2\,424\,833.$$

You may wish to *check* that this is so by constructing the symbol (7.4). You'll need no piece of technology more sophisticated than a hand calculator, though you could program a computer if you wanted to. We found (see [37]) that the symbol with $b = 274\,117$ that begins with $a_1 = 1$ has 19 entries and we have not found any symbol with $b = 274\,177$ having fewer entries. Further, with $b = 2\,424\,833$, the symbol beginning with $a_1 = 1$ has 237 entries, but the symbol beginning with $a_1 = 65\,537$ has only 213 entries. We don't know whether the symbols we have found verifying the factors of F_6 and F_9 are the shortest possible. Can any reader do better? (Notice that $65\,537 = F_4$!)

8

A geometry and algebra thread – Constructing, and using, Jennifer's puzzle

Required materials
- Strips of paper (preferably of different colors)
- Heavy paper, such as lightweight cardboard
- Paper clips

Optional materials
- Ruler
- Compass

8.1 Facts of life

In many instances involving the *use* of geometry in the real world, we need to make adjustments to take into account the *realities of life*. For example, paper comes in various thicknesses (which are never zero!) and the interior of every container must be *larger* than what it contains. These and other very elementary facts of reality affect *how* we are able to take practical advantage of the theorems obtained from our study of the geometry of idealized lines, planes, and solids.

As this section title implies, we concern ourselves here with the details of *practical construction*, in this case of a particular set of nested polyhedra. Namely, we construct an octahedron and 4 tetrahedra that fit inside a larger tetrahedron that, in turn, fits inside a cube. As you will see – assuming that you become actively involved in carrying out these instructions – overcoming the difficulties encountered in using, in a real-life situation, a theory that is perfect in principle is very much a skill of the eyes and hands as well as of the mind.

What follows is first a description of the construction, along with some hints about how to *solve* Jennifer's puzzle; this is followed by some important mathematical consequences to be gained by constructing the puzzle. The details first appeared in [60].

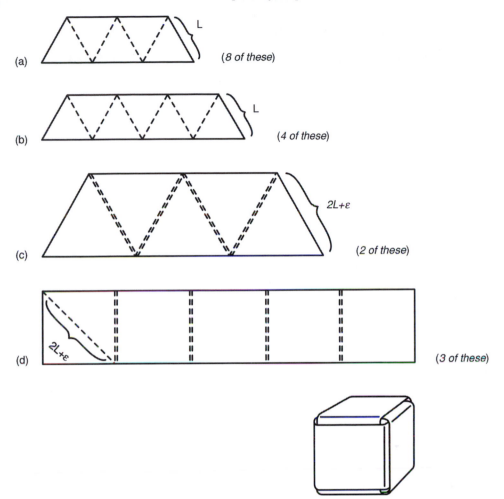

Figure 8.1 Pattern pieces. Note: ϵ here means a small positive number, to be chosen experimentally. Not every occurrence of ϵ need represent the same number.

8.2 Description of the puzzle

This puzzle was created by Jennifer, the daughter of one of the authors (JP), when she was assigned to "make a puzzle based on geometry, and turn it in to the teacher without the solution." It consists of 17 strips of paper and an instruction sheet. Figure 8.1 and the instruction sheet from Jennifer's original puzzle identify the 17 puzzle pieces and, simultaneously, tell what is required in the solution.

Instructions for Jennifer's puzzle: Try it!

1. You get all the little strips of 5 triangles each (there should be 8) and braid them into 4 tetrahedra.
2. Then you get the 4 strips of 7 triangles each and braid an octahedron (that is, an 8-faced polyhedron).
3. Now you take the 2 big strips of 5 triangles each and braid a large tetrahedron as before; but in this one you put the 4 little tetrahedra and the octahedron.
4. Finally, take the 3 strips of 5 squares each and braid a cube, into which you put the large tetrahedron.

GOOD LUCK!

Jennifer Pedersen

9th Grade Geometry Project

Castillero Junior High School

San Jose, California

8.3 How to make the puzzle pieces

The choice of material for the strips shown in Figure 8.1(a) and (b) is not of great importance, as long as the material has enough bulk and crispness to hold a good fold. Of course, the puzzle will be more interesting visually if you use paper of different colors for different strips; that is, use a different color for each of the *two* strips that form a small tetrahedron, and for each of the *four* strips that form the small octahedron.

If you have already constructed the pentagonal dipyramid or triangular dipyramids, or the rotating ring of Chapter 6, you will know how to make the small strips for the four tetrahedra and the octahedron. In this case we leave it to you to improvise.

However, if you prefer to "start from scratch" then, for each pattern piece, start with a strip of paper *longer* than you really need and draw an angle of $\frac{\pi}{3}$, as shown:

$$\frac{\pi}{3}\ (=60°)$$

Then fold the paper down:

Then unfold:

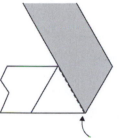

Notice that along the fold line there is a small width (crease) of paper – this crease is usually easier to see on the underside of the paper. The width of the crease will depend on the thickness of the paper used; for this reason it is very important that the next fold line be made *precisely* as shown:

Avoid covering any part of the crease produced by the previous fold line.

Continue folding this way until you have the required number of triangles for the pattern pieces required in steps 1 and 2 of the instructions for Jennifer's puzzle. If you run out of paper, then carefully measure the starting angle on the next strip and continue folding.

The large tetrahedron and the cube are very much more satisfactory when constructed from heavier paper, such as lightweight cardboard. Here again this poses problems along the fold lines. The thicker the paper the more pronounced these difficulties become. What is required is mostly an awareness of the problem so that when you draw the pattern pieces you leave room for the hinge.

For the large tetrahedron, begin by determining, experimentally, the length $2L + \epsilon$, as shown in Figure 8.1(c). This is easily done by placing the completed octahedron and its 4 tetrahedra together, as is shown in Figure 8.2, and *measuring* the length that will be required in order for them to *fit inside* the bigger tetrahedron. Then, using a ruler and compass, or some other method (for example, paper-folding), construct, on a lightweight piece of paper, an equilateral triangle of the appropriate edge length. This *pattern triangle* may then be cut out and used to draw the big strip of 5 equilateral triangles, as shown in Figure 8.1(c). The width of the space between successive triangles may be determined by folding a sample of the heavy paper and measuring the width of the resulting crease. The pattern for the strip is then obtained by first drawing parallel lines so that the distance between them is equal to the height of your pattern triangle. Next, using the pattern triangle, draw the 5 triangles so that each one is separated from its neighbor by a

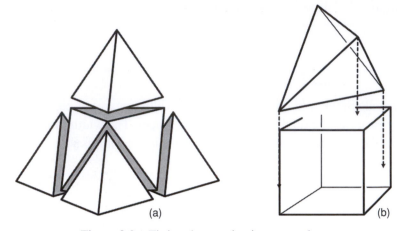

Figure 8.2   Fitting the puzzle pieces together.

pair of parallel lines (providing for the crease between them). Score the crease lines firmly so that, when the strip is cut out, it will fold easily and smoothly between the triangles. Fold each pattern piece so that the score lines will be on the inside of the completed model.

The 3 strips for the cube are constructed using the same underlying principles, with one minor additional feature. Begin by determining, experimentally, the length of the *diagonal* of the required square by placing an edge of the completed big tetrahedron along the diagonal of an oversized square. In this way determine the appropriate size for a *pattern square*. From a piece of lightweight paper, cut out one pattern square. Then draw parallel lines on the heavy paper so that the distance between them is equal to the length of the *side* of the pattern square. Then, using your pattern square, draw the 5 squares of Figure 8.1(d), so that each square is separated from its neighbor by parallel scored lines. But – and this is the additional feature to which we referred – in this case you should make the allowance for the crease about *twice* as wide as that for the width between the triangles. This is because the strips of the cube must wrap *around* each other when the model is constructed (see the diagram of the completed cube in Figure 8.1). These pattern pieces should then be cut out and folded along the score lines so that the score lines will be on the inside of the completed model.

Now, just in case you forget some of the real-life details and end up with pattern pieces that don't fit together nicely because there was not enough allowance made for the hinge, there is a way to salvage your effort. The "way out" does not give as good a result as carefully following our original advice, but it is very comforting psychologically. It is to trim off from each edge of the defective piece a small amount, as shown here:

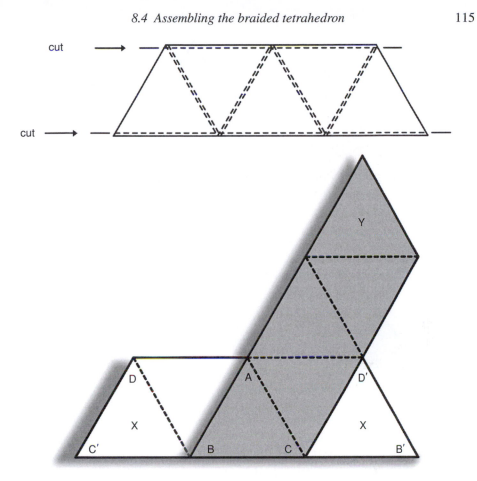

Figure 8.3   Braiding the tetrahedron.

It is not difficult to see that what this does is to truncate at the vertices of the finished model; but if only a small width is trimmed off, the effect on the appearance of the finished model is not noticeable – especially if it is made from brightly colored paper.

Of course, the point of the puzzle is first to figure out how to construct each of the required polyhedra from the specified number of strips and then to get them to fit together as described in the instructions. If you are adventurous, you may wish to try this first on your own. Don't reject this suggestion too quickly; you are very likely to be more successful than you might expect. You can always return to the following instructions later.

8.4 Assembling the braided tetrahedron

On a flat surface, such as a table, lay one strip *over* the other strip exactly as shown in Figure 8.3. The fold lines should all be valley folds as viewed from above the

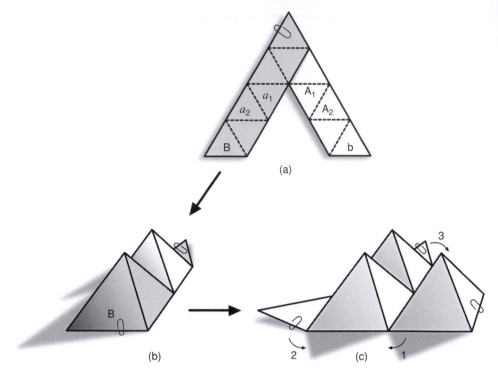

Figure 8.4   Constructing the braided octahedron.

table. Think of triangle ABC as the *base* of the tetrahedron being formed; for the moment, triangle ABC remains on the table. Then fold the bottom strip into a tetrahedron by lifting up the two triangles labeled X and overlapping them, so that C′ meets C, B′ meets B, and D′ meets D. Don't worry about what is happening to the top strip, as long as it stays in contact with the bottom strip where the two triangles originally overlapped. Now you will have a tetrahedron, with three triangles sticking out from one edge. Complete the model by carefully picking up the whole configuration, holding the overlapping triangles X in position, wrapping the protruding strip around 2 faces of the tetrahedron, and tucking the Y triangle into the open slot along the edge BC.

8.5 Assembling the braided octahedron

To construct the octahedron, begin with a pair of overlapping strips held together with a paper clip, as indicated in Figure 8.4(a). Fold these two strips into a double pyramid by placing triangle a_1 under triangle A_1, triangle a_2 under triangle A_2, and triangle b under triangle B. The overlapping triangles b and B are secured with another paper clip, so that the configuration looks like Figure 8.4(b). Repeat this process with the second pair of strips, and place the second pair of braided strips over the first pair, as shown in Figure 8.4(c). When doing this, make certain the flaps with the paper clips are oriented exactly as shown. Complete the octahedron by following the steps indicated by the curved arrows in Figure 8.4(c). You will note that after step 1 you have formed an octahedron; performing step 2 simply places the flap with the paper clip on it against a face of the octahedron; in step 3 you should tuck the flap *inside the model*.

When you become adept at this process you will be able to slip the paper clips off as you perform these last three steps – but this is only an aesthetic consideration, since the clips will be concealed inside the completed model.

After completing the construction of the 4 small tetrahedra and the octahedron, take the models and place a tetrahedron on alternate faces of the octahedron, as shown in Figure 8.2(a). Construct the large tetrahedron containing these polyhedra by placing this configuration on triangle ABC after the two large strips have been put in the position shown in Figure 8.3.

8.6 Assembling the braided cube

The cube may be constructed by first taking one strip and clipping together, with a paper clip, the end squares. Then take a second strip and wrap it around the outside of the "cube" so that one square covers the clipped squares from the first strip and the end squares cover one of the square holes. Secure the end squares of the second strip with a paper clip. Make certain that the overlapping squares of the second strip do not cover any squares from the first strip and that the first paper clip is covered. It should be as shown in Figure 8.5(a). Now slide the third strip underneath the top square so that two squares of the third strip stick out on both the right and left sides of the cube. Tuck the end squares of this third strip inside the model through the slits along the bottom of the cube, as indicated in Figure 8.5(b). If you now turn the completed cube upside down, and place a rubber band around it as shown in Figure 8.5(c) (this is just to make sure the whole thing doesn't come apart when you put the big tetrahedron inside of the cube), the cube may now be opened by pulling up on the strip that covers the top face (this square will be attached inside the model by a paper clip, so you may have to pull firmly) and folding back the flaps that were the last to be tucked inside the model.

If you've done this carefully, you can insert the large tetrahedron into the cube and close the cube by tucking the flaps back into their original positions. At this point you can remove the rubber band. If you have trouble placing the tetrahedron inside the cube, take another look at Figure 8.2(b). Once the tetrahedron is placed inside the cube the paper clips are not necessary for holding the faces of the cube

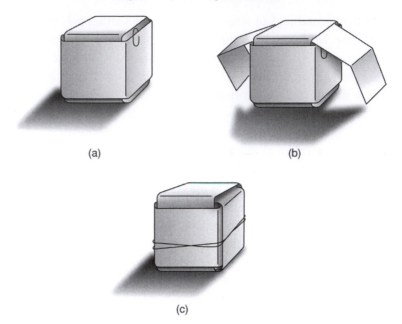

(a)   (b)

(c)

Figure 8.5   Braiding the cube.

in place, because the tetrahedron will exert pressure from inside that will hold the strips in their proper positions.[1]

A Variation

The given construction of the cube (with a different color on each strip) will yield a cube with opposite faces of the same color, because each strip goes alternately over and under each successive strip it meets as it travels around the cube. There is another construction, using the same strips, that produces a cube in which pairs of adjacent faces are the same color. In this construction each strip goes over two strips and then under two strips as it travels around the cube. Once you have mastered the idea of the first construction, you may wish to assemble your cube strips in this alternative configuration.

8.7 Some mathematical applications of Jennifer's puzzle

Volumes of some related polyhedra

We show here how Jennifer's puzzle may be used to calculate the volumes of some Platonic solids. We start from the following three facts about volumes:

---

[1] On most models friction makes the paper clips unnecessary even if the tetrahedron is not inside the cube.

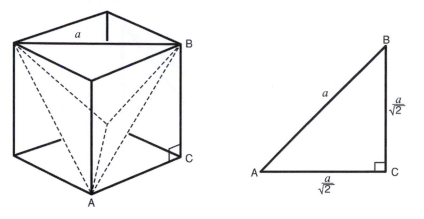

Figure 8.6   A tetrahedron inside a cube.

- The volume of a rectangular parallelepiped of edge-lengths a, b, c is abc.
- The volume of a pyramid of height h standing on a base B is

$$\tfrac{1}{3}h \times (\text{area of } B).$$

- If the linear dimensions of a figure are multiplied by d, the volume is multiplied by d^3.

Now it follows immediately from the first fact that the volume of a cube of edge-length a is a^3. It is perfectly possible to use the second fact to compute the volume of a regular tetrahedron of edge-length a to be $\frac{a^3}{6\sqrt{2}}$. However this calculation requires us to calculate the area of an equilateral triangle of side a – which is $\frac{\sqrt{3}a^2}{4}$; and the height of the regular tetrahedron – which is $\sqrt{\frac{2}{3}}a$. There is, in fact, an easier method provided by thinking "inside the box;" that is, think of a regular tetrahedron placed inside the cube, as in Jennifer's puzzle. The empty space inside the cube and outside the regular tetrahedron then consists of four congruent tetrahedra; and if the regular tetrahedron has edge-length a, the box has edge-length $\frac{a}{\sqrt{2}}$, as can be seen from Figure 8.6. Moreover, each of the four tetrahedra is an upright wedge, as shown in Figure 8.7.

Thus the volume of each wedge is, according to the second fact,

$$\tfrac{1}{3}\,\frac{a}{\sqrt{2}}\Big(\tfrac{1}{2}\big(\tfrac{a}{\sqrt{2}}\big)^2\Big) = \frac{a^3}{12\sqrt{2}}.$$

Since the volume of our original regular tetrahedron (which we call $V_T(a)$) is equal to the volume of the cube minus the volumes of the four wedges, we see that it may be calculated as

$$V_T(a) = \big(\tfrac{a}{\sqrt{2}}\big)^3 - 4\big(\tfrac{a^3}{12\sqrt{2}}\big) = \frac{a^3}{2\sqrt{2}} - \frac{a^3}{3\sqrt{2}} = \frac{a^3}{6\sqrt{2}}.$$

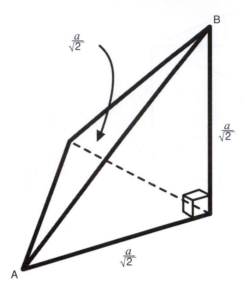

Figure 8.7   A wedge from the cube in Figure 8.6.

Notice that this is the same as the result we claimed earlier. Notice, too, that we have proved the following:

> The volume of a regular tetrahedron is $\frac{1}{3}$ the volume of the smallest cubical box in which it sits.

For, as we calculated, our box has a volume of $\frac{a^3}{2\sqrt{2}}$.

It is now very simple to compute the volume of a regular octahedron of edge-length a. For, as observed in Jennifer's puzzle, we may place regular tetrahedra on four of the faces of the regular octahedron to obtain a regular tetrahedron of edge-length $2a$. Thus if $V_T(a)$ is the volume of a regular tetrahedron (of edge-length a) and $V_O(a)$ is the volume of the regular octahedron (of edge-length a), then

$$V_O(a) + 4V_T(a) = V_T(2a).$$

Then, using the known value for $V_T(a)$, we obtain

$$V_O(a) + 4\frac{a^3}{6\sqrt{2}} = \frac{(2a)^3}{6\sqrt{2}}$$

yielding

$$V_O(a) = \frac{4a^3}{6\sqrt{2}} = \frac{\sqrt{2}\,a^3}{3}.$$

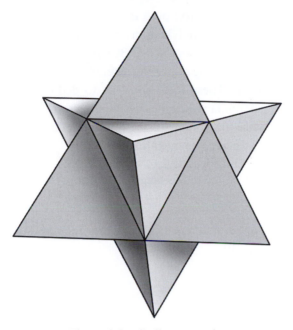

Figure 8.8   Stella octangula.

Notice that this last argument also shows the following:

> The volume of the regular octahedron of edge-length a is 4 times the volume of the regular tetrahedron of edge-length a.

Despite this result, you would be doomed to failure if you tried to *construct* a regular octahedron of edge-length a by putting together four regular tetrahedra of edge-length a – it just can't be done! The volume measures are the same, but that is not enough. We observe, however, that to prove our last result it is quite unnecessary to calculate the volume of the tetrahedron; we apply the formulas for $V_T(a)$ and $V_O(a)$, knowing that by doubling the length of each edge on any polyhedron, the volume of that polyhedron is increased by a factor of 2^3, or 8. And this, of course, we know by virtue of the third fact in our original list of facts about volumes.

However, if you put together the two results displayed (in boxes) earlier, we obtain a third interesting comparison:

> The volume of the regular octahedron of edge-length a is $\frac{1}{2}$ the volume of the regular tetrahedron of edge-length $2a$ and hence $\frac{1}{6}$ the volume of the cube in which the octahedron sits with its vertices at the midpoints of the faces of the cube.

If we glue a regular tetrahedron onto *each* face of a regular octahedron (of edge-length a), we obtain the polyhedron shown in Figure 8.8, called by the great astronomer, Johannes Kepler (1571–1630) the ***stella octangula***. This may be thought of as two interpenetrating regular tetrahedra of edge-length $2a$, intersecting in the regular octahedron. These facts allow us to deduce immediately that the volume of the stella octangula is 12 times the volume of the regular tetrahedron of edge-length a, or 3 times the volume of the regular octahedron of edge-length a. Even more interesting, we have the next result:

> The volume of the stella octangula is $\frac{1}{2}$ that of the smallest cube into which it can be placed.

We will say more about mathematical uses for Jennifer's puzzle (concerning group theory) in Chapter 10.

9

A polyhedral geometry thread – Constructing braided Platonic solids and other woven polyhedra

9.1 A curious fact

In Chapter 8 we gave instructions for braiding tetrahedra, cubes, and octahedra. The natural question to ask is: **Can we construct the other two Platonic solids by a similar braiding technique?** It turns out that we can braid all five of the regular convex solids. In fact, it is easy to verify the following composite statement for the five Platonic solids.

If you make each solid from straight strips of paper fashioned into bands and if you require that all strips on the given model be identical to (or mirror images of) one another, that every edge on the completed model be covered by at least one thickness from the strips, that every face be entirely covered by the strips, and that the *same total area* from each strip must show on the finished model,

Then you can braid

the tetrahedron	from	2 strips
the hexahedron (cube)	from	3 strips
the octahedron	from	4 strips
the icosahedron	from	5 strips
the dodecahedron	from	6 strips.

(The pattern pieces are shown in Figure 9.1.)

We don't have any general explanation for this curious fact, but we will show you how you can easily demonstrate it. This we do by providing you with instructions for constructing the required polyhedra, and once the polyhedra are constructed, you may then verify, simply by looking at them, that they satisfy the conditions of the preceding statement. To establish the conclusion of the statement, all you need to do is take the models apart and count the number of strips for each one – we call this our "proof by destruction." The braided solids are shown in Figure 9.2.

123

124   *A polyhedral geometry thread*

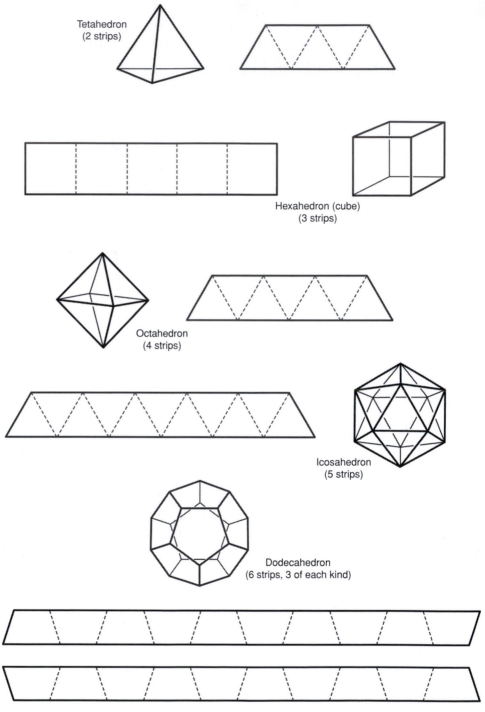

Figure 9.1   Pattern pieces for braided Platonic solids.

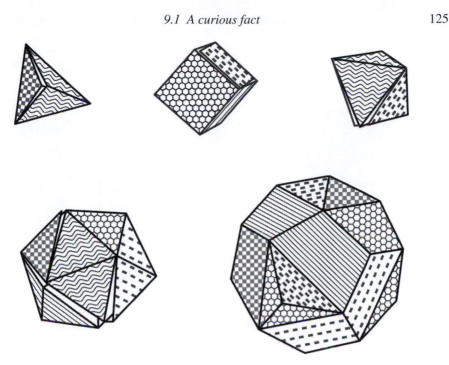

Figure 9.2   The braided Platonic solids.

The instructions for how to braid the tetrahedron, octahedron, and cube are in Sections 8.4, 8.5, and 8.6 respectively. The first thing we will do in the rest of this chapter is to explain the general procedure for preparing the strips with folded gummed mailing tape, and then we will show how to use these strips to braid certain models, including the dodecahedron and icosahedron of Figure 9.2.

However, we feel we owe it to you to emphasize that this is not the *only* way these polyhedra can be braided from straight strips. Thus, for example, a cube may be braided from 4 strips, as seen in Figure 9.3(a); we call this the ***diagonal cube*** (because the edges of the strips lie on the diagonals of its faces) and it is one of the easier models to build. Another example is a dodecahedron that can be braided from 6 strips (such that the resulting model has a pentagonal hole at the center of each face, as seen in Figure 9.3(b). This turns out to be an important model, mathematically, and we call it the ***golden dodecahedron***, because, on the D^2U^2-tape the ratio of the length of the long line to the length of the short line is the same as the ratio of the length of the long line plus the length of the short line to the length of the long line – which is one way to define the ***golden ratio*** (see [78]). The golden dodecahedron is also one of the easier models to construct and is especially satisfying (according to our students).

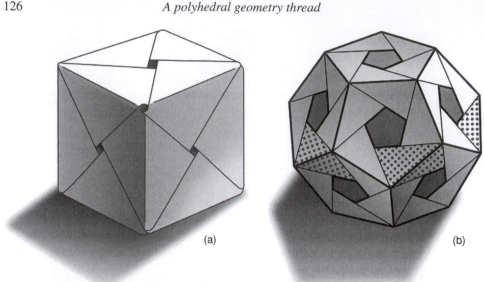

Figure 9.3   (a) The diagonal cube. (b) The golden dodecahedron.

In order to help you improve your construction techniques gradually we give instructions, in the next four sections, for constructing the diagonal cube, the golden dodecahedron, the dodecahedron, and the icosahedron, in that order.

9.2 Preparing the strips

Required materials
- Gummed mailing tape; for a sturdier model, use gummed tape that is reinforced with filament. Tape that is about $1\frac{1}{2}$ to 2 inches (4 to 5 cm) wide is easy for beginners to handle, but if you want larger models, then 3 inch to 4 inch (7 to 10 cm) tape is feasible.
- Paper (preferable colored), onto which you will glue the gummed tape. Butcher paper or gift-wrapping paper is very suitable. Foil wrapping paper can be spectacular, but it sometimes cracks and peels along the creased edges. Experiment with a little piece before making a big investment.
- Scissors
- Sponge (or washcloth)
- Shallow bowl
- Water
- Water bottle, with water in it
- Hand towel (or rag)
- Some heavy books
- Paper clips
- Bobby pins
- Transparent tape (optional)

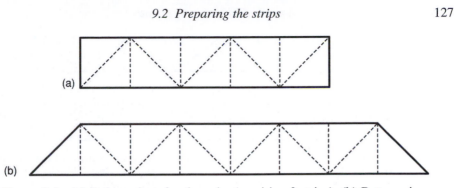

Figure 9.4   (a) Pattern piece for the cube (requiring 3 strips). (b) Pattern piece for the diagonal cube (requiring 4 strips).

Having decided which models you want to build, proceed systematically. Observe that the tetrahedron, octahedron, and icosahedron are the only models in this chapter requiring equilateral triangles. In fact, if you count all the triangles required for these models you will find it is precisely 93 – but it would be tedious and unnecessary to count them as you fold them. It is much easier to simply begin folding the triangles, as in Section 2.3, and then, after throwing away the first few irregular triangles, start cutting off the required pattern pieces; then (when you have just about used up all the folded tape) fold some more triangles, cut off more pieces, . . . until you have all the pattern pieces that require equilateral triangles.

Next construct the pattern pieces for the cube and the diagonal cube. This may be done by using an exact folding process, since by folding the tape directly back on itself you produce an angle of $\frac{\pi}{2}$, and by bisecting that angle you get the angle of $\frac{\pi}{4}$ that will lead across the tape to a point where the tape may be folded back on itself to produce a square. Notice that in order to get the pattern pieces for the cube with 3 strips we only need to use the *short* lines on the folded tape, but the gummed tape for one piece will appear as shown in Figure 9.4(a).

To get the pattern pieces for the diagonal cube (with 4 strips), we only need to use the *long* lines on the folded tape, but the gummed tape will appear as shown in Figure 9.4(b). Take care that the long lines are oriented properly.

Finally, prepare the pattern pieces for the two dodecahedra. Notice that both the dodecahedron of Figure 9.1 and the golden dodecahedron of Figure 9.3(b) require strips that were initially folded by the $U^2 D^2$-(or, equivalently, $D^2 U^2$-) process of Section 2.4.

Notice that, in the pattern pieces for the dodecahedron of Figure 9.1, we only need to use the *short* lines on the $U^2 D^2$-tape, but the gummed tape will appear as shown in Figure 9.5.

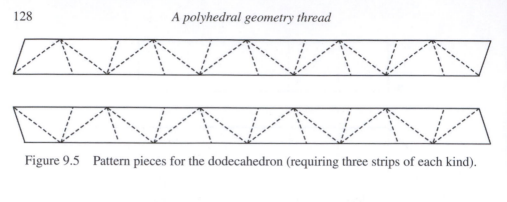

Figure 9.5   Pattern pieces for the dodecahedron (requiring three strips of each kind).

Figure 9.6   Pattern piece for the golden dodecahedron (requiring 6 strips).

Similarly, notice that to obtain the pattern pieces for the golden dodecahedron of Figure 9.3(b), we only need to use the *long* lines on the U^2D^2-tape, but the gummed tape will appear as shown in Figure 9.6. You will lose some triangles between the successive strips if you make all the pattern pieces identical (do you see why?)

Once you have prepared all the gummed tape pieces, get your colored paper and place each of the pieces for a particular model on a different color of paper (to make certain they will all fit). Place a sponge (or washcloth) in a bowl with enough water so that the top of the sponge is very moist or, as we like to say, squishy. Then take each strip (one at a time) and moisten one end by pressing it onto the sponge. Next, holding the moist end, pull the rest of the strip across the sponge. Make certain the entire strip gets wet and then place it on the colored paper. Use a hand towel, or rag, to wipe up the excess moisture and to smooth the tape into contact with the colored paper. (Only now, when the pattern piece has been properly glued to the colored paper, should you begin to worry about cleaning up yourself or the table!)

An efficient scheme is to glue all the pieces that go on one color onto the piece of paper and then go on to the next piece of colored paper. After all the pieces have been glued onto a particular color put some heavy books on top of the pieces of tape so that they will dry flat. The drying process may take several hours – we can't say, even roughly, how long because it depends on your climate!

When the tape is dry, cut out the pattern pieces, trimming off a *very tiny* amount of the gummed tape from the edge as you do so (this serves to make the model look neater and, more importantly, it allows for the increased thickness produced by gluing the strip to another piece of paper). Then refold each pattern piece so that the raised ridges are on the colored side of the paper. For the models involving equilateral triangles you will refold on every line, but it is ***very important*** to remember the following rule:

For the	refold only on the
3-strip cube	SHORT lines
4-strip cube	LONG lines
dodecahedron of Figure 9.1	SHORT lines
golden dodecahedron	LONG lines

Now you are ready to assemble your models. In each case, we strongly urge you first to try it on your own. You may find it useful to clip the beginning and end of the strip with a paper clip, so that they overlap; you can then see how the pieces fit together. As you become an expert you will figure out how to eliminate the paper clips. But if you want more help now, turn to the appropriate section for specific instructions.

9.3 Braiding the diagonal cube

Begin by laying 4 strips on a table with the colored side *down*, exactly as shown in Figure 9.7(b). The first time you do this it may be useful to secure the center (where the 4 strips cross each other) with some transparent tape. Set a water bottle on the center square to hold the strips in place. Now braid the vertical faces around the bottle, remembering that each strip goes alternately over and under the other strips it meets – and every face has four colors on it. Next, fold the last triangle on each strip into a horizontal position and secure the ends with paper clips before removing the water bottle. The arrangement should look like Figure 9.7(c). Complete the top of the cube by removing the paper clips. At that point the four "ends" closest to the center can be flipped into a symmetric position on top of each other. The remaining four "ends" can then be tucked in so that the colors of each strip come into contact with a strip of the same color and the cube will look like Figure 9.7(d). If you can braid this without the transparent tape and the bottle you are an expert!

9.4 Braiding the golden dodecahedron

Begin by taking five of the strips identical to the one shown in Figure 9.8(a) and arrange them as shown in Figure 9.8(b). Secure them in this position with paper

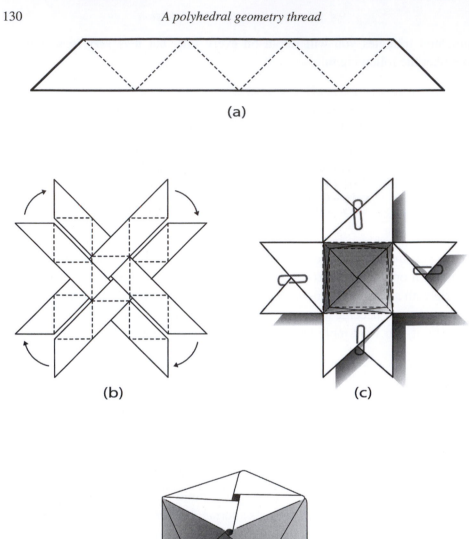

(a)

(b)   (c)

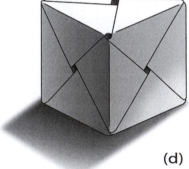

(d)

Figure 9.7   (a) Pattern piece. (b) The beginning layout. (c) The partially completed model. (d) The completed diagonal cube.

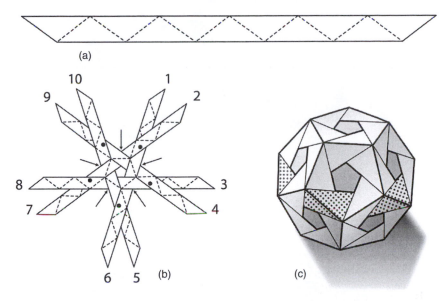

Figure 9.8 Braiding the golden dodecahedron.

clips at the points marked with arrows (note that this paper clip will be in the center of a face, so it won't interfere with any of the edges or vertices). View the center of the configuration as the North Pole. Lift this arrangement and slide the even-numbered ends clockwise over the odd-numbered ends to form the five edges coming south from the Arctic pentagon. Secure the strips with paper clips at the points indicated by the heavy dots. Now weave in the sixth (equilateral) strip, shown patterned in Figure 9.8(c), and continue braiding and clipping, where necessary, until the ends of the first five strips are tucked in securely around the South Pole. During this last phase of the construction it is important to *keep calm* and to *take your time*! Just make certain that every strip goes alternately over and under every other strip all the way around the model. Figure 9.9 shows the construction in progress.

When the model is complete all the paper clips may be removed and the model will remain stable.

9.5 Braiding the dodecahedron

Recall that you will need 6 strips as shown in Figure 9.5. Take 2 of these strips, which are mirror images of each other, and cross them as in Figure 9.10.

Secure the overlapping edge with a bobby pin, and then make a bracelet out of each of the strips in such a way that

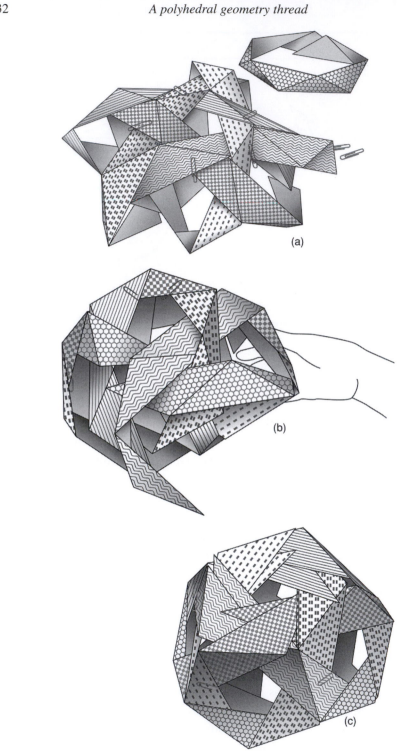

Figure 9.9 Progressive stages of the construction. (a) The "North Pole" is complete with the equatorial strip nearby. (b) The equatorial strip comes together. (c) Five strips at the "South Pole" as they are tucked in.

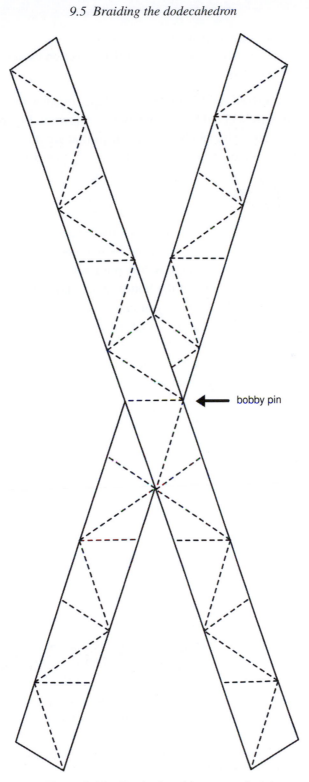

Figure 9.10   Beginning the construction.

(1) four sections of each strip overlap, and
(2) the strip that is *under* on one side of the bracelet is *over* on the opposite side. (This will be true for both strips.)

Use a bobby pin to hold all 4 thicknesses of tape together on the edge that is opposite the one already secured by a bobby pin, as shown in Figure 9.11(a).

Repeat the above steps with another pair of strips that are mirror images of each other. Now you have two identical bracelet-like arrangements. Slip one inside the other one as illustrated in Figure 9.11(b) so that it looks like a dodecahedron with triangular holes on four faces.

Take the last two strips and cross them precisely as you did earlier (reversing the crossing would destroy some of the symmetry); then secure them with a bobby pin on the crossing edge. Carefully put two of the loose ends through the top hole and pull them out the other side so that the bobby pin lands on CD in Figure 9.11(b). Then put the other two ends through the bottom hole and pull them out the other side, as shown in Figure 9.11(c). Now you can tuck in the loose flaps, but make certain to reverse the order of the strips – that is, whichever one was underneath at CD should be on top when you do the final tucking (and, of course, the top strip at CD will be underneath when you do the final tucking).

After you have mastered this construction you may wish to try to construct the model with tricolored faces, shown in Figure 9.12. This construction and the one just described are both very similar to the construction for the cube in Section 8.6. The difference is that in the case of the dodecahedron, the three "bracelets" that are braided together are each composed of two strips. This illustrates, rather vividly, exactly how to inscribe a cube symmetrically inside the dodecahedron. To put it another way, it shows how the dodecahedron may be constructed from the cube by placing a "hip roof" on each of the 6 faces of the cube.

9.6 Braiding the icosahedron[1]

Label one of the 5 strips, as shown in Figure 9.1, with a "1" on each of its 11 triangles; you should write on the side of the paper that will be on the *inside* of the finished model. Then label the next strip with a "2" on each of its triangles, the next with a "3" on each of its triangles, the next with a "4" on its triangles, and, finally, the last with a "5" on its triangles.

Now lay the 5 strips out so that they overlap each other *precisely* as shown in Figure 9.13, making sure that the center 5 triangles form a shallow cup that points away from you. You may need to use some transparent tape, or masking tape (note

---

[1] According to our students this is the hardest model to build, so we don't recommend that you attempt it until you have built up your skill by constructing other models.

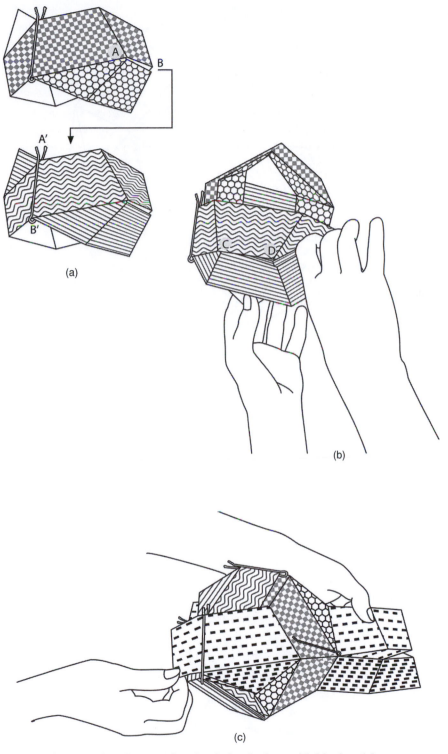

Figure 9.11   Constructing the dodecahedron with bicolored faces.

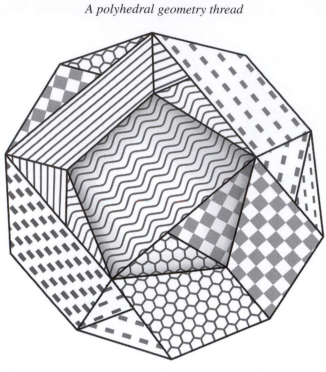

Figure 9.12   A braided dodecahedron with tricolored faces.

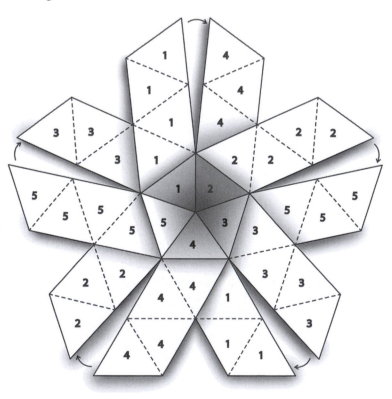

Figure 9.13   The beginning layout for constructing the icosahedron.

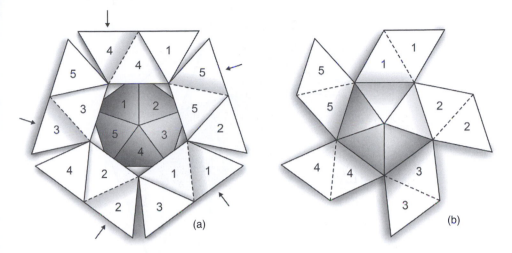

Figure 9.14   Constructing the icosahedron.

that it won't show on the finished model) to hold the strips in this position. If you do need to use the tape, it works best to put it along the 5 lines coming from the center of the figure.

Now study the situation carefully, before making your next move. You must bring the 10 ends up so that the part of the strip at the tail of the arrow goes under the part of the strip at the head of the arrow (this means "underneath" as you look down on the diagram; it is really on the outside of the models your are creating, because we are looking at the inside of the model). Half the strips wrap in a clockwise direction and the other end of each of those strips wraps in a counterclockwise direction. What finally happens is that each strip overlaps itself at the top of the model. But, in the intermediate stage, it will look like Figure 9.14(a). At this point it may be useful to put a rubber band around the emerging polyhedron just below the flaps that are sticking out from the pentagon. Then lift the flaps as indicated by the arrows and bring them toward the center so that they tuck in, as shown in Figure 9.14(b).

Now simply lift flap 1 and smooth it into position. Do the same with flaps 2, 3, and 4. Complete the model by tucking flap 5 into the obvious slot. The vertex of the icosahedron nearest you will look like Figure 9.15.

For mathematical uses of these braided models see Chapters 10 and 14.

9.7 Constructing more symmetric tetrahedra, octahedra, and icosahedra

Part of the charm of the diagonal cube and the golden dodecahedron is that, in both cases, the braided model retains all of its inherent symmetry. Generally speaking,

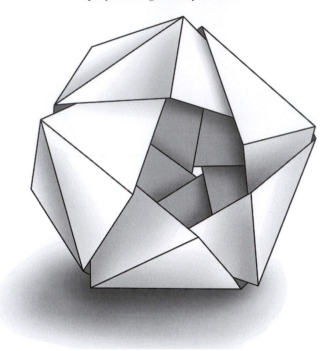

Figure 9.15   The completed braided icosahedron.

braided models lose some of the symmetry of the underlying geometric figure; indeed, our braided tetrahedron, octahedron, and icosahedron all lost some of their underlying symmetry. Thus it is natural to ask: ***Is it possible to braid the tetrahedron, octahedron, and icosahedron in such a way as to retain all the symmetry of the original polyhedron?***

We have, in fact, discovered a way to do this (see [24] or [41]). Figure 9.16(b) shows a typical straight strip of 5 equilateral triangles with a slit in each triangle from the top (or bottom) edge to (just past) the center.[2] The symmetric tetrahedron is constructed beginning with 3 such strips interlaced as shown in Figure 9.16(a). We leave the completion of the model as a challenge to you.

Figure 9.17(a) shows the layout of 3 strips for the beginning of the construction of the symmetric octahedron. We'll give you one more hint. When you use Figure 9.17(a) remember that the strip shown below it in Figure 9.17(b) has to be braided into the figure above it.

---

[2] Theoretically the slit could go just to the center, but the model is then impossible to assemble. You need to leave some leeway for the pieces to be free to move during the process of construction – although they will finally land in a symmetric position so that it looks as though the slit need not have gone past the center of the triangle.

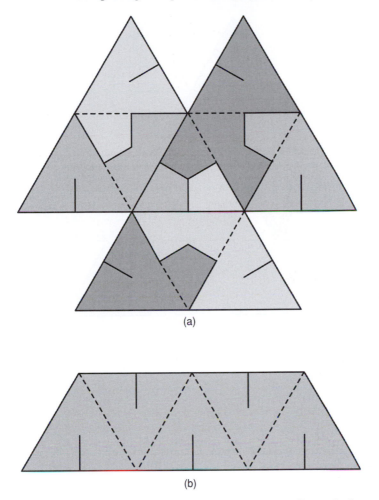

Figure 9.16   Pattern piece and layout for the symmetric tetrahedron.

The symmetric icosahedron may be constructed from 6 strips of this type having 11 triangles on each strip. Over to you! But take heart – these models take several hours to construct. This set of symmetric braided models will look like Figure 9.18 when completed.

9.8 Weaving straight strips on other polyhedral surfaces

Branko Grünbaum and Geoffrey Shephard (see [19–21]) have carefully analyzed certain geometric objects which represent an idealization of woven fabrics in the plane and their investigations lead, among other things, to remarkable theorems concerning the number and nature of the different kinds of what they call

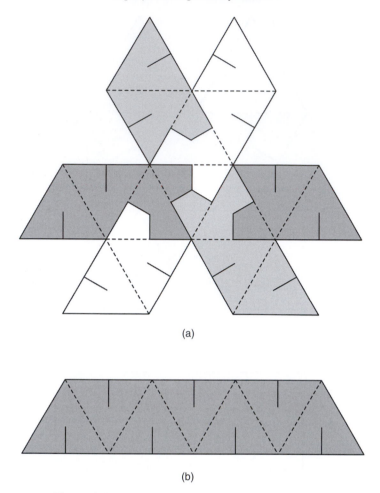

(a)

(b)

Figure 9.17   Layout for the symmetric octahedron.

"isonemal"[3] fabrics in the plane. Their work is fairly exhaustive and they have posed many open problems that may be of interest to our readers. The models shown in this section show the partial results that came from investigating the analog to one of their problems (additional results can be found in [62]).

Figure 9.19 shows an ordinary "over-under" weave in the plane (b) in its idealized form, while part (c) shows how the actual fabric looks, and part (d) shows the layer rankings. Figure 9.20 shows a 3-way, 3-fold weave.

An analog to one of the questions Grünbaum and Shephard asked about isonemal fabrics in the plane would be the following: What is the nature of fabrics woven

---

[3] "Isonemal" is a term from the Greek words ἴσος (the same) and νήμα (a thread or yarn).

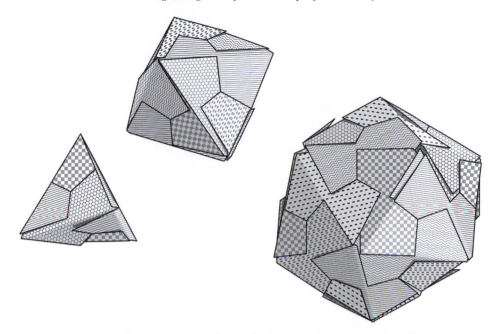

Figure 9.18   Braided symmetric tetrahedron, octahedron, and icosahedron.

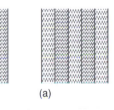

(a)

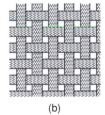

(b)

(c)

21	12	21	12	21
12	21	12	21	12
21	12	21	12	21
12	21	12	21	12
21	12	21	12	21

(d)

Figure 9.19   A 2-way, 2-fold weave represented 3 ways.

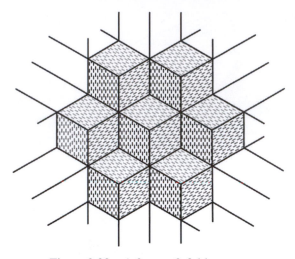

Figure 9.20   A 3-way, 3-fold weave.

on sphere-like polyhedra where you use, for strands, cylindrical-like rings instead
of straight strips (which were used in the plane), and then require that the closures
of the strands completely cover the polyhedron in some specified uniform and
symmetric way? The notion is considered metrically, and the resulting fabrics
are referred to as ***woven polyhedra*** or, more specifically, ***isonemal fabrics on
polyhedral surfaces***.

In the examples shown, the only requirements are that (a) the entire surface of
the model is covered, (b) each face is covered by the same number of straight strips,
and (c) the model retains the symmetry of the underlying structure from which it
came. In particular, there is no requirement of convexity.

One such example is the diagonal cube of Section 9.3. According to the
Grünbaum–Shephard notation, this would be a 4-way, 2-fold isonemal surface.
Suppose you identify each strip with a number 1, 2, 3, and 4. Label each face of
the strip with its number. Figure 9.21 shows one of the possible rankings of the
strips for that model when it is completed.

The model shown in Figure 9.22 is a 10-way, 3-fold isonemal weaving on
the surface of a model with icosahedral symmetry; that is, the underlying surface is that of a dodecahedron where each pentagonal face has been replaced by
5 equilateral triangles. The model is braided from 10 strips each having 19 equilateral triangles (the last triangle is a tab that is glued to the first triangle on
each strip). The model can be made with all the pentagonal pyramids pointing
either outward or inward; if the pyramids point inward the model is remarkably
stable.

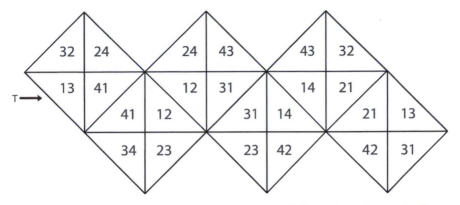

Figure 9.21   Rankings of the 4 strips crossing each face of the diagonal cube.

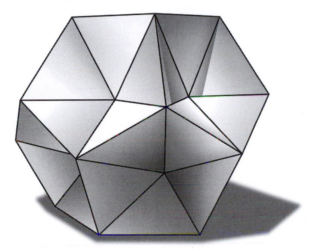

Figure 9.22   The 10-way, 3-fold isonemal polyhedron that is one possible offspring of a dodecahedron.

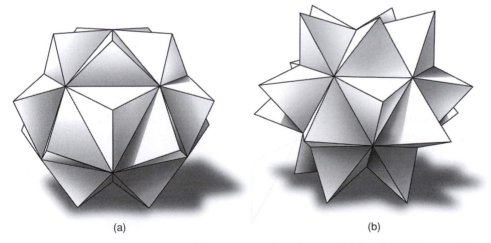

(a)   (b)

Figure 9.23   Two possible offspring of the icosahedron. (a) is a 2-fold isonemal weave and (b) is a 3-fold isonemal weave.

Figure 9.23 shows how one can take an icosahedron and, by adding pyramids to each of its faces, obtain a surface that lends itself to an isonemal weaving. In Figure 9.23(a) the pyramids each have right angles at their apex; the resulting surface allows a 2-fold isonemal covering. In Figure 9.23(b) the pyramids are composed of three equilateral triangles; the resulting surface allows a 3-fold isonemal covering. For more exotic surfaces and instructions on how to construct them see [62].

10

Combinatorial and symmetry threads

10.1 Symmetries of the cube

We first consider the *symmetries* of a cube. After talking (rather a lot) about this important concept, we will go back to Jennifer's puzzle from Chapter 8 to see how it casts light on the relation of the symmetries of a cube to the symmetries of a regular octahedron and those of a regular tetrahedron.

We picture the cube occupying a certain part of space; by a *symmetry* we mean the effect of a rotation of the cube about its center that brings it into a position occupying the same original part of space. Thus, for example, we may rotate the cube through an angle of $\frac{\pi}{2}$ about an axis passing through the midpoints of two opposite faces; this is a symmetry of the cube. It is plain that

1. if we follow one symmetry by another, the composite effect is again a symmetry,
2. if we reverse a symmetry we again get a symmetry, and
3. the "zero" rotation, that is, the "rotation" that holds every point fixed, is trivially a symmetry.

These three facts allow us to talk of the *group* of symmetries of the cube (or, more generally, of the *group* of symmetries of any polyhedron). Notice that a symmetry of a cube is completely determined when we describe the position of the points of the cube after the rotation – it is thus sufficient to describe the destinations of each vertex.

Now, a classical way to study the group of symmetries of the cube is to look at the 4 *main diagonals* of the cube, that is, the 4 straight-line segments that pass from a vertex of the cube to the diametrically opposite vertex. It is plain that any symmetry of the cube permutes these 4 main diagonals, in the sense that, on executing the rotation, some main diagonal (perhaps the original one) comes to occupy the position in space originally occupied by any given main diagonal (see Figure 10.1 for an example).

145

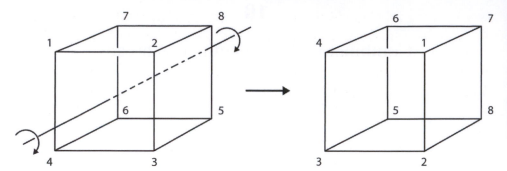

Figure 10.1   Rotation through an angle of $\frac{\pi}{2}$ about the axis joining the midpoints

of front and back faces. Diagonal $\begin{Bmatrix} 15 \\ 26 \\ 37 \\ 48 \end{Bmatrix}$ moves into position originally occupied by

diagonal $\begin{Bmatrix} 26 \\ 37 \\ 48 \\ 15 \end{Bmatrix}$.

The following notation is often used to specify a permutation. We refer to the main diagonals 15, 26, 37, and 48 as D_1, D_2, D_3, and D_4, respectively. Then the permutation described in Figure 10.1 may be written

$$\begin{pmatrix} D_1 & D_2 & D_3 & D_4 \\ D_2 & D_3 & D_4 & D_1 \end{pmatrix}$$

indicating that

D_1 moves into the position originally occupied by D_2;
D_2 moves into the position originally occupied by D_3;

•

•

However, there is an even more convenient notation for this permutation (which we will adopt when we list all the symmetries of the cube).

Let us give another example. The permutation written in the more cumbersome notation as

$$\begin{pmatrix} D_1 & D_2 & D_3 & D_4 \\ D_3 & D_4 & D_1 & D_2 \end{pmatrix}$$

appears, in cyclic notation, as

$$(D_1 D_3)(D_2 D_4)$$

for D_1 "goes to" D_3, which goes to D_1, and D_2 goes to D_4, which goes to D_2. This permutation is thus a composition of 2-cycles, each of length 2. Every permutation can be written in cyclic notation as a composite of cycles. If some element is fixed under the permutation (for example, a rotation about a main diagonal fixes that diagonal), we think of that element as constituting a cycle of length 1. The permutation that moves nothing is called the **identity**; in cyclic notation it is $(D_1)(D_2)(D_3)(D_4)$.

We have seen then that every rotation of the cube induces a permutation of the set of 4 main diagonals. It is less obvious that given any permutation of the 4 main diagonals, there is exactly one symmetry of the cube that effects this permutation. Let us give you just one key argument leading to this important conclusion. Let us ask: **What symmetry could transform each main diagonal into itself?** If such a symmetry sends vertex 1 to vertex 1, we claim it must send vertex 2 to vertex 2. For, if not, it sends vertex 2 to vertex 6, and this is impossible because 12 is an edge of the cube and 16 is not. Likewise it must send vertex 3 to 3 and vertex 4 to 4. In other words it is the zero movement (or rotation). Thus it follows that the only non-trivial symmetry leaving the diagonal alone, if it existed, would have to send vertex 1 to vertex 5, vertex 2 to vertex 6, vertex 3 to vertex 7, and vertex 4 to vertex 8.

We are going to use the idea of *orientation* to show that there is no such symmetry. If we orient the faces of the cube so that *opposite* orientations are induced in the common edge of two faces – this is called **orienting the cube** – and if we orient the face 1234 by $\overrightarrow{1234}$, then we must orient the face 2358 by $\overleftarrow{2358}$, so that we induce opposite orientations in their common edge 23. Likewise we must orient the face 5678 by $\overleftarrow{5678}$. Figure 10.2 shows this orientation of the cube (one of two possible orientations). Precisely it shows the faces of the cube in Figure 10.1 drawn in net form with the orientation of each face indicated by the orientation of any one face. It is now plain that 1234 cannot be moved by a symmetry to 5678, since a rotation must preserve orientation, so that there is *no* non-trivial symmetry leaving the 4 main diagonals alone. A consequence of this is that distinct symmetries must produce distinct permutations of the main diagonals.

It is not difficult to see that there are precisely $4! = 4 \cdot 3 \cdot 2 \cdot 1 = 24$ permutations of 4 objects. And there are also precisely 24 symmetries of the cube! You should experiment with one of the cubes you have constructed. We suggest that you first use it to verify that a cube really does have 24 symmetries. One way to do this is to suppose you have a cube with each face colored a different color – for the sake of discussion, say the colors are red opposite blue, green opposite orange, and white opposite purple. Further, suppose your cube lives in an imaginary cubical drawer that is only slight larger than the cube. Figure 10.3 shows how it might look when you open the drawer.

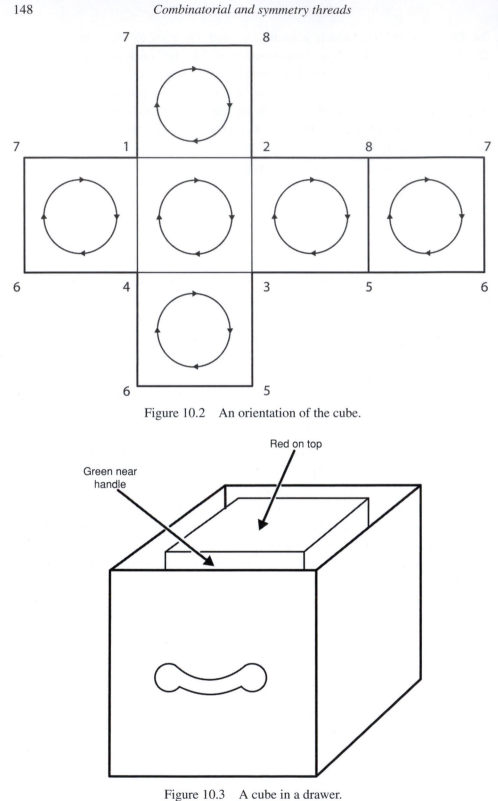

Figure 10.2   An orientation of the cube.

Figure 10.3   A cube in a drawer.

Table 10.1.

Description of rotation about the axis through the centers of	The amount of rotation	The number of rotations of this type
Opposite faces	$\pm\frac{1}{4}$ turn	6
Opposite faces	$\frac{1}{2}$ turn	3
Opposite vertices	$\pm\frac{1}{3}$ turn	8
Opposite edges	$\frac{1}{2}$ turn	6
Identity	0	1
		Total number of rotations = 24

Now if you want a particular face, say red, to show at the top of the drawer, then there are exactly 4 ways you can achieve this, since you can have green, orange, white, or purple against the handle of the drawer. (Why can't blue be against the handle of the drawer?) Once you have decided which color you want at the top of the drawer and which color is against the handle, the position of the cube will be completely determined. Do you see why? Since we may choose the top face in 6 ways and then choose the front face in 4 ways, it follows that the cube can be set down in the drawer in exactly 24 ways.

It now follows that the symmetries that produce the 24 possible positions of the cube in the drawer really do correspond to the 24 permutations of the 4 main diagonals of the cube. You will find it very instructive to study Table 10.1 which describes each symmetry geometrically. The very important zero rotation, which is also called the *identity*, is listed last in the table. Naturally, it induces, the *identity* permutation of the vertices and the main diagonals. The force of our earlier arguments is that the group of symmetries of the cube is *isomorphic* to (that is, structurally equivalent to) the group of permutations of 4 objects, usually written S_4, and called *the permutation group of 4 objects*.

Table 10.2 contains even more information about symmetry. For example, if in addition to the data about how the vertices and the main diagonals are permuted, you record the number and length of the various types of cycles in these permutations, you may be able to observe a very interesting pattern that is part of a bigger picture.

10.2 Symmetries of the regular octahedron and regular tetrahedron

There is a very nice geometrical argument that now allows us to determine the symmetries of the regular octahedron; for, as you will recall from Jennifer's puzzle, in Chapter 8, a regular octahedron can fit inside a cube with its vertices at the

Table 10.2.

Geometric description of the type of symmetry, where the numbers 1, 2, . . . , 8 refer to vertices as labeled in Figure 10.4	Type of permutation of the vertices	Type of permutation of the main diagonals
Type 1 A $90°(= \frac{\pi}{2})$ rotation in either direction about the axis through the centers of opposite faces; for example, through the centers of the front and back faces. (There are 6 of these; find the other 5.)	Two 4-cycles. In our example, $(1234)(5678)$	One 4-cycle. In our example $(D_1 D_2 D_3 D_4)$
Type 2[1] A $180°(= \pi)$ rotation through the centers of opposite faces; for example, through the centers of the front and back faces in one direction. (There are 3 of these; find the other 2.)	Four 2-cycles. In our example, (13) (24) (57) (68)	Two 2-cycles. In our example, $(D_1 D_3)$ $(D_2 D_4)$
Type 3 A $120°(= \frac{2\pi}{3})$ rotation in either direction about the axis through opposite vertices; for example, through vertices 1 and 5 in one direction. (There are 8 of these; find the other 7.)	Two 3-cycles and two 1-cycles. In our example, (247) (368) (1) (5)	One 3-cycle and one 1-cycle. In our example, $(D_2 D_4 D_3)$ (D_1)
Type 4 A $180°(= \pi)$ rotation in either direction about the axis through the centers of opposite edges; for example, through the centers of edges 12 and 65. (There are 6 of these; find the other 5.)	Four 2-cycles. In our example, (12) (73) (84) (56)	One 2-cycle and two 1-cycles. In our example, $(D_1 D_2)$ (D_3) (D_4)
Type 5 No movement. (This is called the *identity* or *zero rotation*.) (Of course, there is just one of this type.)	Eight 1-cycles	Four 1-cycles

[1] Normally cycles are written horizontally next to each other, as in Type 1. In this case, and in the following two cases, we use the vertical format merely for typographical reasons.

midpoints of the faces of the cube. Thus every symmetry of the cube will send the octahedron into the space it originally occupied within the cube.

But it is also true that the midpoints of the faces of a regular octahedron are the vertices of a cube. Thus every symmetry of the octahedron will send this cube into the space it originally occupied within the octahedron.

It therefore follows that the symmetry group of the regular octahedron is the same as (or is isomorphic to) that of the cube. In fact, the full group of all permutations of n objects is called the *symmetric group of n objects* and is written S_n. Thus, as we have shown, the group of symmetries of the cube and hence also of the octahedron is S_4, which is, of course, a group having 24 elements. The group S_4 is also known as the *octahedral* group.

It is particularly revealing to relate the symmetries of the cube to the *diagonal cube* of Section 9.3; for that cube is constructed from 4 strips of paper – and these strips are the objects that are permuted by each rotation of the cube! Thus the abstract notion of the permutation of the 4 main diagonals becomes much more vivid when you think of the 4 strips that form the surface of the diagonal cube as the objects being permuted.

The diagonal cube has some combinatorial surprises

Suppose we consider a diagonal cube braided from 4 strips, each a different color. *Let us ask in how many ways four colors can be arranged in a circle.* The answer is 6, since any of the four colors can be placed at the top of the circle and the other three can then be placed in 3! ways. Moreover, our diagonal cube presents all 6 of these arrangements on its faces, with opposite faces displaying opposite arrangements. Suppose we now ask *in how many ways four colors can be arranged three at a time in a circle.* The answer is 8, since we can eliminate any one of the four colors, and the remaining 3 colors can then be arranged in a circle in 2 ways. Then, our diagonal cube presents all 8 of these arrangements at its vertices, with opposing vertices displaying opposite arrangements. In this case the missing color creates an "equatorial" strip about the cube such that if you drew a line along the center of that strip you would have a perfect hexagon. Thus we easily see that every cube has a regular hexagonal cross-section. Finally, *let us ask in how many ways you can choose 2 colors from a collection of 4 colors.* The answer is 12 since the first color may be any of the four colors and the second any of the remaining 3. And, our diagonal cube has 12 edges, each of which is crossed by two of the strips, with the order of the 2 strips reversed on opposite edges.

Now let us again return to Jennifer's puzzle. In particular, let us suppose you place the big tetrahedron inside the cube with the vertices of the tetrahedron at the vertices 2, 4, 5, 7 of the cube in Figure 10.4. Then consider the symmetries of the

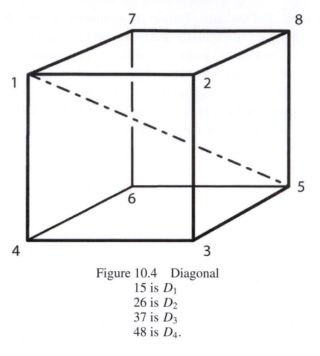

Figure 10.4   Diagonal
15 is D_1
26 is D_2
37 is D_3
48 is D_4.

cube. We claim the following: A symmetry of the cube *either* permutes the vertices 2, 4, 5, 7 among themselves, or it moves them into positions originally occupied by 1, 3, 6, 8. Moreover, exactly half of the symmetries of the cube fall into the first class and exactly half into the second. You can verify these facts in a number of ways. If all else fails, use the symmetries in Table 10.2.

It now follows that the group of symmetries of the regular tetrahedron is precisely the same as the subgroup of symmetries of the cube consisting of those symmetries that lie in the *first* class described earlier. As we pointed out, this group has half of the elements of S_4, that is, 12 elements. In fact, this is the subgroup of S_4 called the *alternating group*, written A_4 (see Courant and Robbins [7]) for further details. It is also known, particularly among geometers, as the *tetrahedral* group.

An equivalent way to identify the subgroup A_4 is to ask the question: **Which permutations of the main diagonals of the cube move vertex 2 to one of the vertices 2, 4, 5, 7 and which move it to one of the vertices 1, 3, 6, 8?** Once again, those in the first class constitute the symmetries of the regular tetrahedron.

Remark on orientation and symmetry

In our study of the symmetries of the cube, you will remember that we used a rather sophisticated argument about the orientation of the cube (to show that no symmetry of the cube could send each vertex of the cube to the diametrically opposite vertex). We would like to clarify this argument for you by describing the

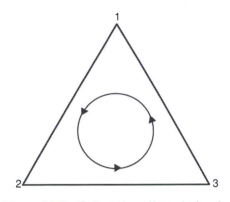

Figure 10.5   Oriented equilateral triangle.

analogous situation for the symmetries of an equilateral triangle oriented as shown in Figure 10.5.

If we allow only *planar* symmetries of the triangle (that is, rotations that take place in the plane of the triangle), then all we can do is to rotate through $\pm 120°(= \pm\frac{2\pi}{3})$ about the center of the triangle – or, of course, to carry out the zero rotation. Thus the only permutations (in our first cumbersome notation) of the vertices are the following three:

$$\begin{pmatrix} 1 & 2 & 3 \\ 1 & 2 & 3 \end{pmatrix} \qquad \begin{pmatrix} 1 & 2 & 3 \\ 2 & 3 & 1 \end{pmatrix} \qquad \begin{pmatrix} 1 & 2 & 3 \\ 3 & 1 & 2 \end{pmatrix}.$$

It is obvious that these all maintain the orientation of the triangle. However, if we allow a rotation in space (of 3 dimensions), then by rotation through $180°(= \pi)$ about the line joining vertex 1 to the midpoint of side 23, we may achieve the permutation

$$\begin{pmatrix} 1 & 2 & 3 \\ 1 & 3 & 2 \end{pmatrix}$$

which, as you may observe, *reverses* the orientation of the triangle. Thus in talking about the symmetries of the regular 3-gon (and a similar remark applies to the symmetries of the regular *n*-gon), we should specify whether we insist on planar symmetries or allow rotations in 3-dimensional space.

Now comes the crucial point! In talking about the symmetries of the cube, we should also specify whether we confine our rotations to the 3-dimensional space of the cube (as we have, in fact, done) or allow rotations in a 4-dimensional space! In the former case, orientation is preserved, but it is not necessarily preserved in the latter. It is a tribute to our awareness of the 3-dimensional world in which we live that we adopt (mathematically) different conventions in discussing symmetries of

Table 10.3.

	V	E	F
Tetrahedron	4	6	4
Hexahedron (cube)	8	12	6
Octahedron	6	12	8
Dodecahedron	20	30	12
Icosahedron	12	30	20

planar and non-planar figures. In fact, by rotation in 4-dimensional space, we can achieve the apparently impossible symmetry of the cube described in this section. Do you see how?

10.3 Euler's formula and Descartes' angular deficiency

Let us look at the five Platonic solids and record, for each of them, the number of vertices (V), edges (E), and faces (F). The result is shown in Table 10.3.

We notice that, in all cases, we have the formula

$$V - E + F = 2. \tag{10.1}$$

This formula is called ***Euler's Formula*** for polyhedra. Euler produced arguments to show that the formula (10.1) holds for *any* polyhedron in our sense (as defined in Chapter 5). For example, for the pentagonal dipyramid, we have

$$V = 7, \qquad E = 15, \qquad F = 10, \qquad \text{and} \qquad 7 - 15 + 10 = 2,$$

as promised by Euler's Formula.

We will not prove the formula here – but we will show you a modified version of an argument due to George Pólya (1887–1985), that Euler's Formula is equivalent to a very deep result, due to the great French mathematician and philosopher of the seventeenth century, René Descartes (1595–1650).

Consider any of the convex polyhedra whose construction is described in this book. If you consider all the faces that come together at a particular vertex and lay them out flat, they will leave a gap. Thus, for example, for the regular tetrahedron we would get, at any vertex, the gap shown in Figure 10.6(a), which is π. For the cube we would get, at any vertex, the gap shown in Figure 10.6(b), which is $\frac{\pi}{2}$. In fact, it was Euclid who pointed out (see Section 5.3) that there would always be a (positive) gap for any convex polyhedron. Descartes called this gap the ***angular deficiency*** of the polyhedron at the particular vertex. Let us number the vertices of a given polyhedron and write δ_n for the angular deficiency of the polyhedron at the

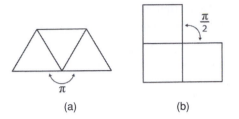

Figure 10.6   (a) The angular deficiency for a vertex of the tetrahedron. (b) The angular deficiency for a vertex of the cube.

nth vertex. Descartes studied the **total angular deficiency** of a convex polyhedron, that is, the sum of the angular deficiencies at each vertex. Let us write Δ for the total angular deficiency, so that

$$\Delta = \delta_1 + \delta_2 + \cdots + \delta_V = \sum_{n=1}^{V} \delta_n.$$

He proved the remarkable fact that, for *any* convex polyhedron,

$$\delta = 4\pi.$$

Again, we will not attempt to prove this, but we will follow Pólya's line of reasoning to show that $V - E + F = 2$ and $\Delta = 4\pi$ are equivalent statements. In the course of doing so, we obtain a result that takes us far beyond the domain of convex polyhedra as we defined them in Chapter 5. In fact, what we will prove is that

$$\Delta = 2\pi(V - E + F)$$

and you should immediately see that this identity establishes the equivalence of $V - E + F = 2$ and $\Delta = 4\pi$. However, we will make no use of convexity in our argument, and we will not need to assume that our polyhedron is deformable into the surface of a sphere; it will suffice that it is constructed out of polygonal faces, where 2 faces are put together by "gluing" a side of one to a side of the other to form an edge of the resulting surface. This last condition has the following important consequence. Let S be the total number of sides of the faces of our surface; then

$$S = 2E.$$

For example, a tetrahedron consists of 4 triangles and each triangle has 3 sides; thus $S = 12$, while $E = 6$. Or, for the dodecahedron, we have 12 pentagons and each pentagon has 5 sides; thus $S = 60$, while $E = 30$.

We are now ready to prove that $\Delta = 2\pi(V - E + F)$. What we do is to count the *sum of the face angles* (which we call A) in two different ways. We first count by vertices. Now, since the angular deficiency at the nth vertex is δ_n, the sum of the face angles at the nth vertex is $2\pi - \delta_n$. Thus

$$A = \sum_{n=1}^{V}(2\pi - \delta_n) = 2\pi V - \Delta. \tag{10.2}$$

We next count by faces. Now, if a face has m sides, then the sum of the interior angles is $(m - 2)\pi$, since, as we showed in Section 3.1, the sum of the *exterior* angles is 2π. Thus, if our polyhedron has F_m m-gons among its faces, those m-gons contribute $(m - 2)F_m\pi$ to the sum of the face angles. We thus arrive at the key formula

$$A = F_3\pi + 2F_4\pi + 3F_5\pi + \cdots = \sum_m (m - 2)F_m\pi = \left(\sum_m mF_m - 2\sum_m F_m\right)\pi.$$

Now

$$F = F_3 + F_4 + F_5 + \cdots = \sum_m F_m.$$

Also, each m-gon has m sides, so that the contribution to the number of sides from the m-gons is mF_m. Thus

$$S = 3F_3 + 4F_4 + 5F_5 + \cdots = \sum_m mF_m.$$

We put together these last three formulas, along with the fundamental relationship $S = 2E$, to infer that

$$A = (S - 2F)\pi = (2E - 2F)\pi.$$

Comparing this with (10.2), the earlier formula for A, obtained by counting by vertices, we conclude that

$$2\pi V - \Delta = 2\pi(E - F)$$

or

$$\Delta = 2\pi(V - E + F) \tag{10.3}$$

as claimed.

We repeat that the result (10.3) is very general and takes us well beyond the very restricted class of (convex) polyhedra that are discussed in this book. Of course, we have to allow *negative* angular deficiencies if we no longer insist that our polyhedra be convex.

An interesting non-convex example is furnished by the 12-celled collapsoids of Section 12.3. You may count the constituent parts to see for yourself that $V = 26$, $E = 72$, $F = 48$, so that $V - E + F = 2$. The different kinds of deficiencies are listed next:

12 vertices, surrounded by 4 equilateral triangles, contribute an angular deficiency of $\frac{2\pi}{3}$ each;

8 vertices, surrounded by 6 equilateral triangles, contribute an angular deficiency of 0 each; and

6 vertices, surrounded by 8 equilateral triangles, contribute an angular deficiency of $-\frac{2\pi}{3}$ each (that is, an angular excess of $\frac{2\pi}{3}$).

Thus we see that the sum of all the angular deficiencies for the 12-celled collapsoid is

$$12\left(\tfrac{2\pi}{3}\right) + 8(0) + 6\left(-\tfrac{2\pi}{3}\right) = 6\left(\tfrac{2\pi}{3}\right) = 4\pi$$

so that (10.3) holds in this non-convex case, too, as promised.

Another interesting example is furnished by our rotating ring of tetrahedra as seen in Section 6.4. We imagine each of the "linking edges" between adjacent tetrahedra which enable us to rotate the ring, pulled apart into two edges; this is necessary in order to retain the relationship $S = 2E$. If our ring is made up of k regular tetrahedra, then you may verify that

$$V = 2k, \qquad E = 6k, \qquad F = 4k$$

so that $V - E + F = 0$. On the other hand, 6 equilateral triangles come together at *every* vertex, so that the angular deficiency at each vertex is 0. Again, we see that for the rotating ring of tetrahedra the formula (10.3) is splendidly vindicated, this time in a case in which the surface is definitely *not* deformable into a sphere.

In fact, the configuration we get when we pull apart the linking edges is what we would call a ***rectilinear model of a torus*** (like a *bicycle tire*). That is to say, just as our Platonic solids, if made out of a malleable material, could be deformed into the shape of a sphere, so could our rotating ring be deformed into a torus, usually depicted as shown in Figure 10.7.

We now come to an interesting point about the relationship (10.3), which we hope will appeal to our readers. First, the quantity $(V - E + F)$ is called the ***Euler***

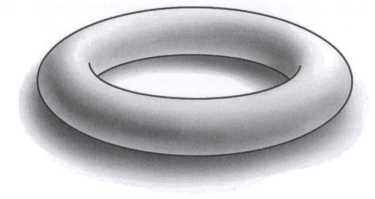

Figure 10.7   A torus.

characteristic, and denoted by χ, so that (10.3) would be expressed as

$$\Delta = 2\pi \chi. \tag{10.4}$$

As we have already observed for our sphere-like polyhedra, χ had the value of 2, and for toroidal polyhedra it would appear, from our example of the rotating ring, that χ has the value of 0. It is not just a matter of the two sides of our relationship (10.4) being equal; the Euler characteristic is obviously a *combinatorial invariant*, that is, it depends on the way our configuration is broken up into faces, edges, and vertices. In fact, we know that it depends on far less – for example, the value is *always* 2 provided only that the configuration can be deformed into a sphere. It is an example of a *topological* invariant (see [7] and [31]). It is, at any rate, obvious that the quantity χ is unaltered if we distort the polyhedron somewhat (for example, if we massage a cube so that the faces are simply quadrilaterals). It is, however, by no means obvious that Δ is unaffected by such a distortion, since the angular deficiency at any particular vertex would certainly be expected to undergo change. Thus (10.4) tells us that, although Δ is defined by means of certain angular measures, it is in reality independent of those measures and depends only on the topological type of the polyhedron.

10.4 Some combinatorial properties of polyhedra

We continue here to think of a polyhedron in the more general sense considered in the previous section; such a polyhedron may be described as a *closed, rectilinear surface*. Thus, as before, our surface has V vertices, E edges, F faces, S sides, and F_m m-gonal faces, so that

$$F = \sum_m F_m, \qquad S = \sum_m m F_m, \qquad S = 2E.$$

We now prove two further basic relationships. They are so important that we call them theorems.

Theorem 10.1 $2E \geq 3F$; *equality holds if and only if each face is a triangle.*

Theorem 10.2 $2E \geq 3V$; *equality holds if and only if exactly 3 faces come together at each vertex.*

Before proving these theorems, we invite you to verify them for the Platonic solids, using Table 10.3. You can also verify them for the convex deltahedra discussed in Section 5.2, and for the collapsoids discussed in Chapter 12.

Proving Theorem 10.1 is very easy in view of the relationships we gave just before its statement. Remember that, in forming the sums $\sum_m F_m$ and $\sum_m m F_m$, the number m takes values $3, 4, 5, \ldots$ Thus

$$\sum_m m F_m \geq \sum_m 3 F_m, \quad \text{equality holding if and only if } F_4 = F_5 = \cdots = 0.$$

It follows that $S \geq 3F$, equality holding if and only if each face is a triangle. Since $S = 2E$, Theorem 10.1 is proved.

Our argument suggests that, to prove Theorem 10.2, we want to break up the vertices in a way analogous to that in which we classified the faces into m-gons for various m. Thus we write V_m for the number of vertices at which m faces come together, and we then want to prove that

$$V = \sum_m V_m, \quad 2E = \sum_m m V_m.$$

Given these equalities, Theorem 10.2 is proved just as we proved Theorem 10.1. Since the first of these equalities is easily seen to be true, we concentrate on the second. We first remark that if m *faces* come together at a vertex, then m *edges* come together at that vertex (indeed, the analogy with the earlier classification of faces would perhaps have been better illustrated by talking of the number of edges coming together at a vertex rather than the number of faces). Thus if we count by vertices, we count in all $\sum_m m V_m$ edges, but each edge is counted twice since an edge joins 2 vertices. Thus

$$\sum_m m V_m = 2E$$

as claimed. If we want an analog of the idea of a side as used in Theorem 10.1, it is that of a *ray*, emanating from a given vertex. If R is the number of rays, then

$$\sum_m m V_m = R = 2E.$$

Table 10.4 *Pairs of dual polyhedra.*

n		V	E	F	n		V	E	F
3		5	9	6	3		6	9	5
4		6	12	8	4		8	12	6
5		7	15	10	5		10	15	7
⋮		⋮	⋮	⋮	⋮		⋮	⋮	⋮
n		$n+2$	$3n$	$2n$	n		$2n$	$3n$	$n+2$

But, although $R = S$, there is no sense in trying to think of each ray as a side or each side as a ray.

The proof of Theorem 10.2 is now easily completed (compare the proof of Theorem 10.1). We have

$$2E = \sum_m mV_m \geq 3 \sum_m V_m = 3V$$

and equality holds if and only if $V_4 = V_5 = \cdots = 0$, that is, if and only if exactly 3 faces come together at every vertex.

In comparing the proofs of these two theorems, we find ourselves on the threshold of an exciting idea, that of a polyhedron and its *dual*. This is a pairing of polyhedra such that, if P and Q are dual polyhedra, then

$$V_m(P) = F_m(Q), \qquad F_m(P) = V_m(Q), \qquad E(P) = E(Q).$$

For example, the cube and the octahedron are dual; so, too, are the dodecahedron and icosahedron. The tetrahedron is dual to – itself! Actually, the duality is richer than we have indicated, but we have probably said enough! For more about this idea and its applications see Section 15.4.

The duality of the cube and the octahedron is a special case of the duality between a dipyramid having an n-gon for its equator and the prism having n-gons for bases, as shown in Table 10.4. Likewise, the self-duality of the tetrahedron is a special case of the fact that every pyramid having an n-gon for a base is self-dual, as shown in Table 10.5. You may wish to check these statements from the data in

Table 10.5 *Self-dual polyhedra.*

n		V	E	F
3		4	6	4
4		5	8	5
5		6	10	6
⋮		⋮	⋮	⋮
n		$n+1$	$2n$	$n+1$

the tables. We haven't done all the work for you; you should work out the values of F_m and V_m, for various m, yourself.

Let us close this section by drawing your attention to certain very concrete consequences of our theorems.

Corollary 10.3 *If all faces on a surface are triangles, then the number of faces is even and the number of edges is divisible by* 3.

Corollary 10.4 *If* 3 *faces of a surface come together at each vertex, then the number of vertices is even and the number of edges is divisible by* 3.

Corollary 10.5 *A polyhedron (in the strict sense) cannot have* 7 *edges.*

We will be content to prove the third corollary, confidently leaving the proofs of the other two corollaries to you (remember that they are consequences of Theorems 10.1 and 10.2).

To prove Corollary 10.5, we suppose that $E = 7$ and hope that this will lead to a contradiction. Since $2E \geq 3F$, we have $3F \leq 14$, so that $F \leq 4$; similarly (using Theorem 10.2 instead of 10.1), $V \leq 4$. But then

$$V - E + F \leq 4 - 7 + 4 = 1$$

contradicting Euler's Formula, $V - E + F = 2$, for a polyhedron. Corollary 10.5 is proved.

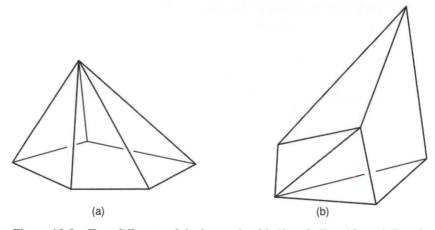

(a) (b)

Figure 10.8 Two *different* polyhedra, each with $V = 6$, $E = 10$, and $F = 6$.

Suppose you try the same argument with $E = 10$; we again assume we have a polyhedron in the original, strict sense. We have $3F \leq 20$, so that $F \leq 6$. Then

$$V - E + F \leq 6 - 10 + 6 = 2.$$

Since, in fact, $V - E + F = 2$, we must have $F = 6$, and $V = 6$. From the equations

$$3F_3 + 4F_4 + 5F_5 + 6F_6 \cdots = 20$$
$$F_3 + F_4 + F_5 + F_6 \cdots = 6$$

we infer (subtracting 3 times the second equation from the first) that

$$F_4 + 2F_5 + 3F_6 + \cdots = 2.$$

Thus $F_6 = F_7 = \cdots = 0$ and we have just two possibilities:

$$F_4 = 0, \qquad F_5 = 1, \qquad \text{giving} \quad F_3 = 5$$

or

$$F_4 = 2, \qquad F_5 = 0, \qquad \text{giving} \quad F_3 = 4.$$

The former possibility is realized by a pentagonal pyramid (in Figure 10.8(a)) and the latter by the polyhedron shown on the right (in Figure 10.8(b)). Note that the two polyhedra in Figure 10.8 are each self-dual, but they are not the same.

11

Some golden threads – Constructing more dodecahedra

11.1 How can there be more dodecahedra?

It's an interesting fact that if you draw a regular pentagon and then extend its sides you will get a regular pentagram (5-pointed star polygon) that surrounds it. Then if you join the vertices of the pentagram you get another pentagon whose extended sides produce yet another pentagram, and so on, each configuration being larger than the previous one. Or, you can go the other way; by beginning with a pentagon you can construct, by joining every other vertex, a pentagram whose sides intersect on the boundary of a smaller regular pentagon. Then the process can be repeated, producing alternately a pentagram, pentagon, pentagram, pentagon, . . . each inside the previous drawing. See Figure 11.1 where the *faces* of the special dodecahedra we will describe in this section are labeled.

Hermann Weyl (1885–1955) recalled that when a pentagram has one vertex pointing straight down it is a symbol for evil and when one vertex is pointing straight up it is a symbol for good. It is amusing that in Weyl's book [81], the pentagram is shown with one side parallel to the vertical side of the page, so that it points neither up nor down. We used the symbol for good in our illustration of Figure 11.1. An interesting feature of any pentagon is that the ratio of its diagonal to its side is the golden ratio.[1] This also means that the ratio of the length of the long line to the length of the short line on the D^2U^2-tape that we folded in Chapter 2 is the golden ratio.

The construction of each dodecahedron described in this chapter involves the use of gummed mailing tape. In each case the tape must first be folded by the D^2U^2- (or U^2D^2-) procedure described in Section 2.4 that produces what we call $\frac{\pi}{5}$-tape. All the models in this section will have 12 faces, and each face will have 5 sides. It is somewhat surprising that 12 of the regular convex pentagons in Figure 11.2(a) can interpenetrate each other to form the *great dodecahedron* – but

---

[1] See Section 9.1 for a defintion of the golden ratio.

163

The faces of the:

  dodecahedron

  small stellated dodecahedron

  great dodecahedron

  great stellated dodecahedron

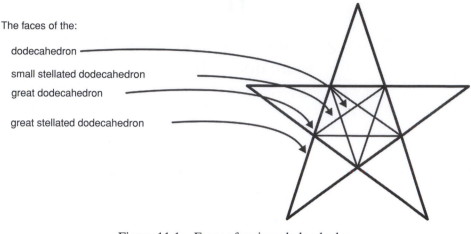

Figure 11.1   Faces of various dodecahedra.

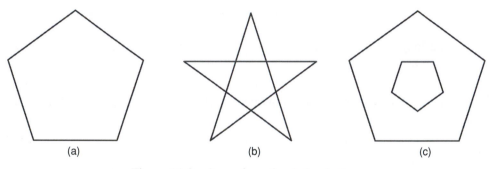

(a)   (b)   (c)

Figure 11.2   Some faces for dodecahedra.

it is even more astonishing that 12 of the *pentagrams* shown in Figure 11.2(b) can
also interpenetrate each other in two very different ways, to form, in one case,
the *small stellated dodecahedron* and, in the other, the *great stellated dodecahedron*. It isn't at all surprising that 12 pentagons with a hole in each, as shown in
Figure 11.2(c) can form the framework of an ordinary convex dodecahedron – but,
in fact, it is very interesting to construct this model, as we have already described
in Section 9.4, by braiding together six strips of the D^2U^2-tape. The results of
these constructions are very beautiful! However, we should point out that, strictly
speaking, these fancy dodecahedra are not polyhedra in the precise sense of the
definition of a polyhedron we gave in Section 5.1. Of course, geometers sometimes
use a less restrictive definition according to which of these models would qualify.

   In this chapter we describe, in detail, how to construct each of these models. In
each case you will need to know the folding procedure of Section 2.4. We refer
to the particular folded tape (the U^2D^2- or D^2U^2-tape) as a $\frac{\pi}{5}$-tape, because the
smallest angle on the tape is $\frac{\pi}{5}$ radians. You may wish to refresh your memory by

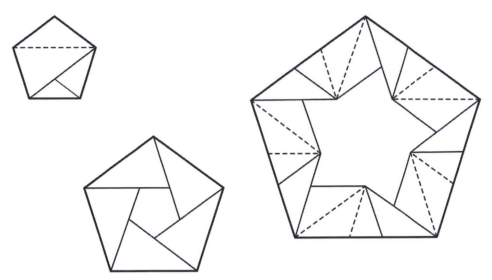

Figure 11.3   Short-line, long-line, and FAT pentagons from the $\frac{\pi}{5}$-tape.

folding a long strip of this tape and creasing it to produce each of the pentagons shown in Figure 11.3.

The dodecahedra for which instructions are given in this chapter are arranged in order of increasing difficulty of construction. In each case it is assumed that you have an ample supply of gummed $\frac{\pi}{5}$-tape.

Before we give you the instructions we list the required materials for this chapter (not all materials are required for each model).

Required materials
- About 9 yards (9 meters) of 2 inch (5 cm) wide gummed mailing tape folded $U^2 D^2$ (to produce $\frac{\pi}{5}$-tape).
- Some gummed 2 inch (5 cm) wide mailing tape that is *not* folded.
- Scissors
- Shallow bowl of water with a sponge (or rag), for moistening the gummed tape when necessary.
- White glue, or glue stick (for paper)
- Ruler
- Some good books (not merely heavy)
- Colored paper (optional)

11.2 The small stellated dodecahedron

Instructions

We first construct a base dodecahedron and then glue a pentagonal pyramid onto each face. The dodecahedron constructed from tape, as described in the "alternative

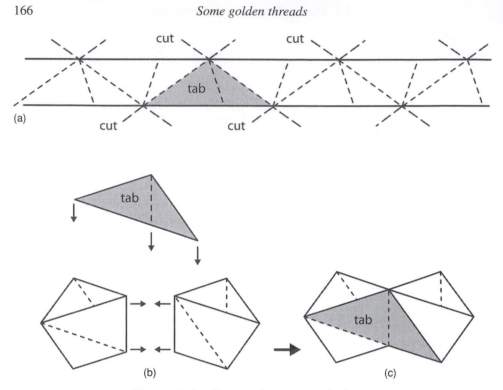

Figure 11.4   Constructing a dodecahedron.

construction" part of Section 5.2, would serve as a suitable base. However, if you do not wish to use those instructions, you may construct the base dodecahedron by another method, which we now describe.

First construct 12 of the short-line pentagons shown in Figure 11.3. As each pentagon is constructed, glue all the *overlapping* portions in place. Use a sponge to moisten the appropriate portion of the tape, bending back the parts that are to remain dry so that just the desired parts of the tape come in contact with the sponge. As each pentagon is completed, put it under (or between the pages of) a large book so that it will dry flat.

While the pentagons are drying, cut 30 tabs from the $\frac{\pi}{5}$-tape as shown in Figure 11.4(a). Notice that all the cuts take place along *long* fold lines of the $\frac{\pi}{5}$-tape. When the pentagons are dry, begin the assembly by taking two pentagons and a tab, as shown in Figure 11.4(b), and glue the tab across the two pentagons to form one edge of the dodecahedron, as shown in Figure 11.4(c). Complete this phase of the construction by continuing to glue pentagons onto free sides of the existing model (so that there are exactly 3 pentagons around each vertex). When all 12 pentagons have been glued in place, part of the construction will be complete. *Practical advice:* When you glue on the last pentagon, it is good to proceed by

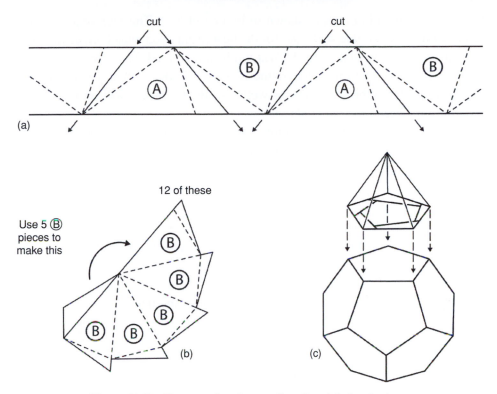

Figure 11.5 Constructing the small stellated dodecahedron.

gluing alternate sides into position around the pentagon (rather than consecutive sides); by doing this, any imperfections in the construction will be more uniformly distributed on the surface of the model (and hence less noticeable).

The final step is to add the stellations to each face of this dodecahedron. Begin this phase by cutting through the small triangles on the $\frac{\pi}{5}$-tape as shown in Figure 11.5(a) to form 60 pieces.[2] Observe that these pieces are not all alike. The Ⓐ pieces and the Ⓑ pieces, as shown in Figure 11.5(a), are *mirror images* of each other. A pentagonal pyramid may be made by gluing together five Ⓑ pieces as shown in Figure 11.5(b). When doing this, bend back the tab portion so that you can press just that part against the sponge. Glue the pieces together so that the gummed side of the tape will be inside the finished model. Recrease each of the fold lines along the hinges before joining the last edge (as indicated by the curved arrow of Figure 11.5(b)). Then the tabs around the base of this pentagonal pyramid may be glued to form a platform around the base of the pyramid. The pyramid

---

[2] The angle is not crucial and some deviation is easily tolerated here. Just make a cut that roughly *bisects* the angle at the vertex of the triangle through which you are cutting.

may then be glued in place, as shown in Figure 11.5(c). This step requires some strong white glue, or glue stick, and for the best results you must hold the pyramid in position until it is well bonded. (You might like to read from one of the "good books" while you wait for each pyramid to dry!) Of course, pentagonal pyramids may be made from the Ⓐ pieces as well, and the 60 pieces will provide all the parts for the 12 required pyramids.

The model may be colored by gluing colored pieces of paper to its faces. These pieces may be prepared by first cutting strips of colored paper of the same width as the tape used for the construction and folding $\frac{\pi}{5}$-strips from which the desired 60 triangles may be cut and glued onto the visible faces of the model.[3] Some craft paper is available with a gummed backing and is particularly suitable for this purpose. A very attractive coloring is achieved by making all the faces that lie on parallel planes of the same color. This, of course, requires exactly 6 colors.

11.3 The great stellated dodecahedron

Instructions

We first construct an icosahedron and then glue a tall triangular pyramid onto each face. Since the base icosahedron must be constructed from equilateral triangles having an edge length equal to the short line on the $\frac{\pi}{5}$-tape, it is necessary to trim off a small amount from the edge of the gummed tape before using it to construct the icosahedron. (Do you see why?) This may be done by beginning to fold the untrimmed tape to produce equilateral triangles (see Section 2.3) and then placing a *short* fold line from the $\frac{\pi}{5}$-tape along one fold line of this tape, thereby determining how much needs to be trimmed off (see Figure 11.6). Mark the tape with a ruler and trim off the necessary amount for about 1 yard (1 meter) of tape. Continue folding the equilateral triangles on this trimmed tape. Use this tape to construct the icosahedron according to the instructions given next.

Referring to Figure 11.7, first cut a section of 11 equilateral triangles. Glue the first triangle over the last to form the equatorial zone of the icosahedron (with the gummed side of the tape on the *inside*). Set that aside and cut a strip of 8 triangles. Take that strip and fold along the lines (1) and (2), shown in Figure 11.7(b), to obtain the hexagon shown in Figure 11.7(c). Glue the overlapping triangles into place and then glue the tab over the triangle next to it, as indicated by the arrow in Figure 11.7(c), thus producing the baseless pentagonal pyramid shown in Figure 11.7(d). Then cut five 2-triangle tabs, one of which is shown in Figure 11.7(e), and glue one end of each piece onto a face of the pyramid. This

---

[3] Save the remaining pieces, since they can be used to cover the visible faces of the great dodecahedron.

Trim this off and continue folding triangles.

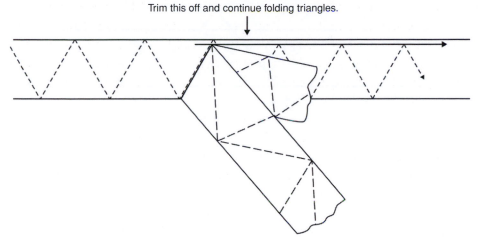

Figure 11.6   Determining the appropriate width of tape to use for constructing the icosahedron.

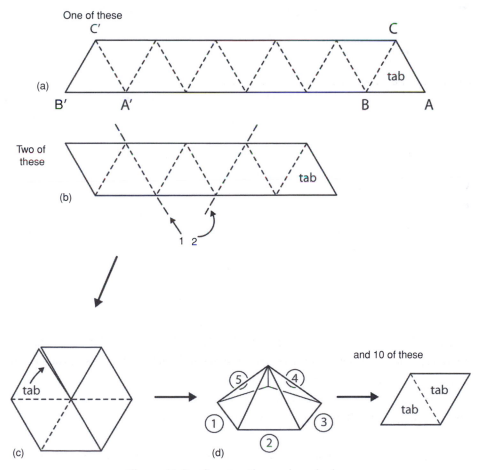

Figure 11.7   Constructing an icosahedron.

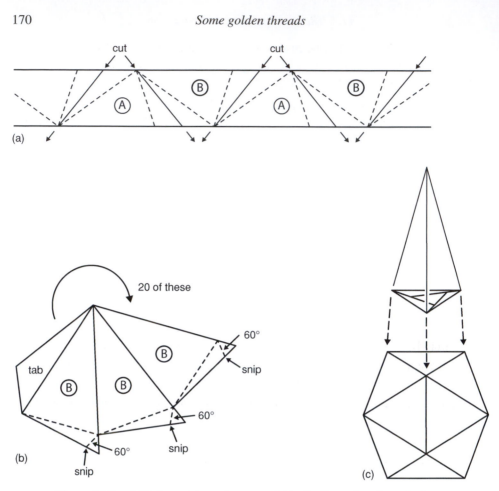

Figure 11.8   Affixing a triangular pyramid to the face of an icosahedron.

pyramid will form the northern (Arctic) region of the icosahedron, and it is attached to the equatorial region by gluing into position, in the order designated, the tabs extending from the edges labeled ①, ③, ⑤, ②, and ④. By following this procedure any imperfections (if there are some) will be distributed around the entire model. The southern (Antarctic) region is completed in precisely the same way. In fact, you can simply rotate the model so as to exchange the North and South Poles and repeat the process of covering the Arctic region!

The final step is to add a triangular pyramid to each face of this icosahedron. Begin, as for the small stellated dodecahedron, by cutting through small triangles on the $\frac{\pi}{5}$-tape, as shown in Figure 11.8(a). As we have already noted, these pieces will be either Ⓐ pieces or Ⓑ pieces. A triangular pyramid may be made by either gluing together three Ⓑ pieces, as shown in Figure 11.8(b), or by gluing together three Ⓐ pieces to obtain a mirror image of Figure 11.8(b). In either case, after joining the last edge of the pyramid (as indicated by the large arrow in Figure 11.8(b)), snip off the excess from the tabs surrounding the base and glue them into position, as shown in Figure 11.8(c). Each of the 20 pyramids may then be glued into place, one at a time, using a good white glue, or glue stick. You will get a better result if you hold each pyramid in position until it is well bonded. (Relax and read some of the good book!)

The model may be colored by gluing pieces of paper onto its faces by the same procedure used with the small dodecahedron. Don't throw away the small colored triangles – you will find them perfect for coloring the faces of the great dodecahedron!

11.4 The great dodecahedron

Instructions

This construction is based on the fact that the visible surface of the great dodecahedron may be obtained by replacing each triangular face of the icosahedron by a particular triangular pyramid that points toward the center of the icosahedron.

The construction of 20 triangular pyramids from the $\frac{\pi}{5}$-tape is initiated by first cutting along *short* transversals to produce 60 sections, as indicated in Figure 11.9(a). The sections produced will be of two types, Ⓐ and Ⓑ, which are mirror images of each other. Separate the pieces into Ⓐ piles and Ⓑ piles. For each of the 60 pieces make an extra fold, bisecting one angle of the larger triangle, as indicated in Figure 11.9(b). Make certain each piece is creased so that it has one valley fold and one mountain fold, exactly as indicated in Figure 11.9(c). The triangular pyramids are formed by taking three Ⓐ pieces (or three Ⓑ pieces) and gluing them together to form a baseless triangular pyramid with tabs protruding from each of its three base edges as shown in Figure 11.9(d). It is important to note that as you look at Figure 11.9(d), the apex of the pyramid is pointing *away* from you and the gummed side of the tape should also be on the side that is not visible.

The construction of the great dodecahedron is now completed by gluing these triangular pyramids together in such a way that (1) the apex of the pyramid points to the center of the polyhedron, and (2) the base edges of the pyramids form the edges of a regular icosahedron. You will note as you do this that at every edge there is a choice of placing the tab *on top* or *underneath*. Don't let this worry you; it makes no difference which way you do it. Just proceed *calmly*, remembering that there should be exactly 5 triangular pyramids around every vertex of the icosahedron. This model is very satisfying to make and you may be surprised at how easily it goes together.

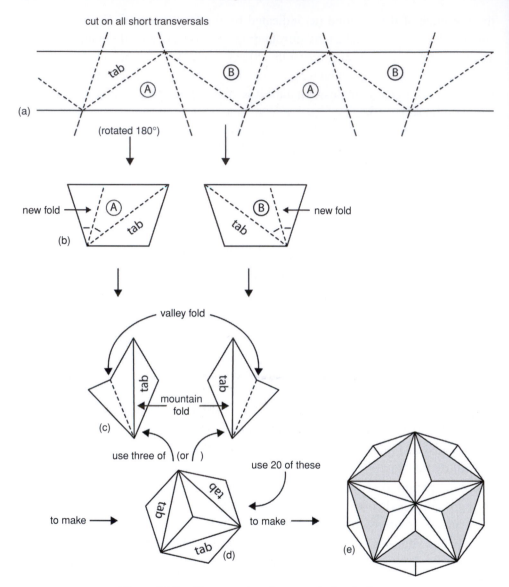

Figure 11.9   Constructing the great dodecahedron.

The completed model, shown in Figure 11.9(e) may be colored by gluing colored pieces of paper onto its faces. In fact, if you have already colored either the small stellated dodecahedron or the great stellated dodecahedron, and if you saved the pieces you did not use for those models, you will find that they are precisely the pieces you need to color this model. A particularly nice coloring is produced when you color the faces that lie in parallel planes the same color.

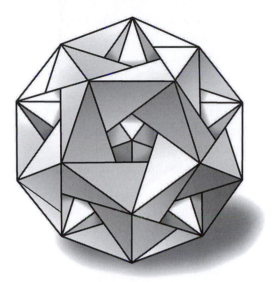

Figure 11.10   The small stellated dodecahedron inside the golden dodecahedron.

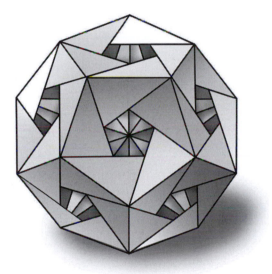

Figure 11.11   The great stellated dodecahedron inside the golden dodecahedron.

11.5 Magical relationships between special dodecahedra

It is a curious fact that the visible faces on the golden dodecahedron, the small stellated dodecahedron, and the great stellated dodecahedron are each composed of exactly 60 isosceles triangles. Furthermore, if all three of these dodecahedra are constructed from tape of the same width (so that the 60 triangles are all the *same size*), then the small stellated dodecahedron fits inside the golden dodecahedron with the stellations protruding through the holes, touching at the midpoints of the edges of those holes (see Figure 11.10).

Furthermore, the great stellated dodecahedron fits *entirely* inside the golden dodecahedron with the vertices of both polyhedra coinciding (see Figure 11.11). The entire arrangement may be colored so that the color of the strips of the golden dodecahedron coincide with the colors on the pentagonal faces of the great stellated dodecahedron. This amazing fact led to the results in Chapter 14 about the extended face planes of the Platonic solids.

Is it any wonder that the pentagon and pentagram are associated with magic?

12

More combinatorial threads – Collapsoids

12.1 What is a collapsoid?

There is an interesting class of polyhedra having the property that all faces are congruent parallelograms. Since all faces are parallelograms, the polyhedra in this class have the property that every edge determines a *zone* of faces such that each face in the zone has two sides parallel to the given edge. Polyhedra having this latter property are called **zonohedra**; we may speak of an *n*-zonohedron to emphasize that the polyhedron in question has *n* zones. As interesting examples of polyhedra in this class, the *rhombic dodecahedron* (which has 12 faces and 4 zones) and the *rhombic triacontahedron* (which has 30 faces and 6 zones) appear in Figure 12.1, which is based on illustrations by H. S. M. Coxeter (1907–2003).

In [8], Coxeter describes the general theory of zonohedra and states that the angles on the faces of the rhombic dodecahedron are $70°32'$ and $109°28'$, while the angles on the faces of the rhombic triacontahedron are $63°26'$ and $116°34'$. You will readily believe that these angles are *not* ones that we get easily by folding paper (though we could get them in principle!). However, these are beautiful models and we can construct polyhedra like them by replacing each of the rhombic faces with a 4-faced pyramid without its base, which is composed of 4 equilateral triangles. We call this a **cell** and refer to each of the triangles as a ***triangular face*** of the cell. One of the authors (JP) experimented with such cells in the hope that the flexibility of the cell might make it possible to approximate the rhombic faces and thereby make it possible to study the symmetry of this model (see [57]).

The experiment showed that the desired models cannot be made with each pyramid projecting out from the polyhedron's center. However, when each pyramid projects in toward the center of the polyhedron, you get a *pseudo-zonohedron*. Furthermore, the models turn out to have a very surprising property apparently not possessed by the real zonohedra, namely, they *fold up and lie flat*. (But as we will see, the original zonohedra can also be made to fold up and lie flat under certain

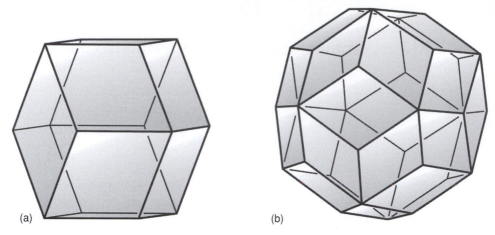

Figure 12.1   (a) Rhombic dodecahedron ($n = 4$). (b) Rhombic triacontahedron
($n = 6$).

circumstances!) All we have to do is leave unattached a sequence of edges that go
from any vertex to the vertex diametrically opposite it. This very surprising and
pleasing feature was first discovered by JP's son (Chris Pedersen, then 9 years old)
while he was playing with these models (as his mother was preparing supper).

Because these pseudo-zonohedra can fold up in various ways, we have named
them *collapsoids – polar* if they collapse about an axis between two poles, and
equatorial if they collapse about an equatorial zone. In the pages that follow we
give you step-by-step instructions for constructing and collapsing these models.
Then in Sections 12.7 and 12.8 we suggest some investigations you might want to
make for yourselves.

12.2 Preparing the cells, tabs, and flaps

Required materials

- Gummed mailing tape; for sturdier models, use gummed tape that is reinforced with
  filament. Any width between $1\frac{1}{2}$ and 3 inches (4 and 8 cm) will be easy to handle.
- Scissors
- Sponge (or washcloth)
- Shallow bowl
- Water
- Hand towel (or rag)
- Colored paper, preferably with self-adhesive backing (optional)

Begin by taking the gummed tape and folding a strip of 50 or more equilateral
triangles (as shown in Section 2.3). Leave the folded tape attached to the roll so

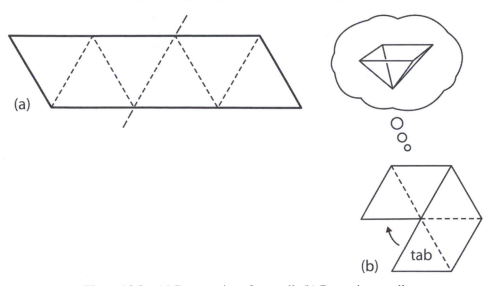

Figure 12.2   (a) Pattern piece for a cell. (b) Becoming a cell.

that you can fold more triangles as you need them. Observe that the new triangles you fold will become more and more accurate as long as you don't cut off the last triangle and start again from scratch. Remember to cut off and discard the first few irregular triangles. Once you have the process started you can cut off from the tape the number of triangles required to construct the cells, tabs, and flaps.

Each of the 4 collapsoids discussed in this chapter requires a certain number of these *cells*, *tabs*, and *flaps*, which are described next. Table 12.1 at the end of this section tells you precisely how many cells, tabs, and flaps are required for each of the 4 collapsoids whose construction is outlined in the following sections. We suggest that you look through the sections of this chapter and decide which model, or models, you would like to make, then construct all of the required parts, and finally turn to the directions that tell you how to glue those parts together.

Cells

Each cell is constructed from a straight strip of 6 equilateral triangles, as shown in Figure 12.2(a). Fold this strip, as indicated in Figure 12.2(b), and glue the overlapping portions together (if the sticky sides are not together, fold the paper in the other direction).[1] It should look like the bottom part of Figure 12.2(b). As the bubble in Figure 12.2(b) indicates, this piece really aspires to be a "baseless" pyramid. To achieve such a pyramid, overlap the triangle labeled *tab* with the

---

[1] Moistening the required triangles may be done by patting the gummed side of each required triangle against a moist sponge.

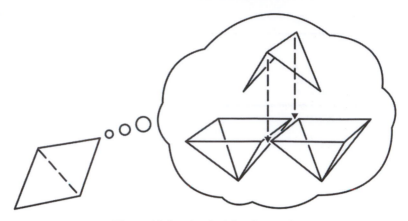

Figure 12.3   A tab doing its work.

triangle indicated by the arrow. Once you see that your result looks like that shown in the bubble, glue the overlapping triangles together.

We call this baseless pyramid a ***cell***. Notice that each cell may be pressed flat in two directions. As you make each cell lay it flat, first in one direction and then in the other, and while it is flat, crease the two edges very firmly. Then place the cells on a table to dry. You may wish to stack them on top of each other in piles that make it easy for you to keep track of how many you have so far constructed (say 5 or 6 to a pile).

Tabs

Tabs, those pairs of triangles that will be used to connect the cells together, are the easiest parts to construct – you simply cut off sections containing two triangles each.

A word of caution is required here, however. We should remind you that, because of the way the tab fits on the completed model, the hinge should be creased so that the sticky sides of the tape come together (see Figure 12.3).

When you have made the tabs, stack them in a pile.

Flaps

The purpose of the flap, which is a tab that only gets attached to a cell by one of its triangles, is to hold the model together when it is expanded. Flaps allow us to open up some edges of the polyhedron and fold it flat for storing. Since the flap will stay in place only if it is fairly stiff, we need to make it more sturdy. Here is one way to do this. Begin with a 3-triangle piece of tape and a 2-triangle piece of tape. Glue one triangle of the 2-triangle piece to the center triangle of the 3-triangle

Table 12.1.

Collapsoid	Number of cells	Number of tabs	Number of flaps
12-celled polar	12	20	4
20-celled polar	20	35	5
30-celled polar	30	54	6
12-celled equatorial	12	18	6

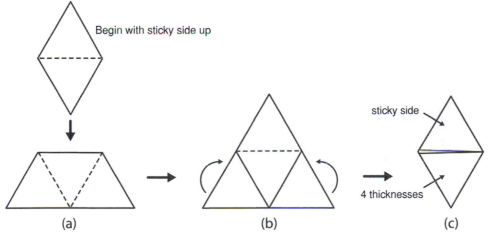

(a)   (b)   (c)

Figure 12.4   Constructing a flap.

piece as shown in Figure 12.4(a). Make certain the sticky sides of both pieces of the tape are facing you. Then wrap the end triangles of the 3-triangle piece over the center triangle, as shown in Figure 12.4(b), and glue them in place, as shown in Figure 12.4(c). Press the 4 thicknesses flat and crease the remaining edge firmly. Stack the flaps in a pile separate from the tabs.

Table 12.1 tells how many of the various constituent parts are required for the four collapsoids explicitly discussed in this chapter.

12.3 Constructing a 12-celled polar collapsoid

Figure 12.5(a) shows a typical cell. Students have found it helpful to put the numbers 1, 2, ..., 12 on the cells in the net diagram of Figure 12.5(c), in any order. Then number your own 12 cells with the numbers 1, 2, ..., 12 before beginning the construction. This helps you to see the connectivity as you construct the model. In the case of 20 or 30-celled collapsoids you do the same thing using, of course, numbers that go from 1 to 20, or 1 to 30, respectively.

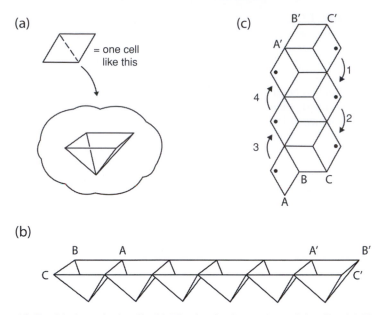

Figure 12.5   (a) A typical cell. (b) The beginning string of 6 cells. (c) How the string is modified to obtain the "net-like" arrangement, ready for tabs.

Next begin the construction by joining 6 cells together in a string that looks like Figure 12.5(b). Then attach to the sides BB′ and CC′ the remaining 6 cells so that the arrangement looks precisely like Figure 12.5(c) *as you look down on it*. At this point it may help you to label the triangles at the head and tail of each arrow with an identifying number (as suggested by the numbers next to the curved arrows). Then use the tabs to join the sides of the cells with the same numbers. The vertices of the cells labeled with heavy dots will then be next to each other. You will notice that an edge going from one heavy dot A(A′) to B(B′) to C(C′) and ending at the other heavy dot will remain open. Your model should now look like Figure 12.6.

At this point we suggest you try collapsing your model. It should fold flat in the shape of $\frac{4}{6}$ of a regular hexagon. Press it fairly gently into the flattened position and bring it back to its expanded shape several times so that you get the feel of the mechanical motion. Now all that remains is to attach the flaps.

Flaps should be attached to provide a covering for the 4 open edges. One very effective way is to attach flaps alternately to one or other of the loose sides along the open edge. More precisely, think of the edges as labeled 1, 2, 3, and 4 as you traverse from North Pole to South Pole on this model; then attach flaps to the left-hand side on edges 1 and 3 and to the right-hand side on edges 2 and 4. The effect of this is that the flaps *interlock* and hold the model together better than they would if all the flaps had been attached to the same side along the open edge.

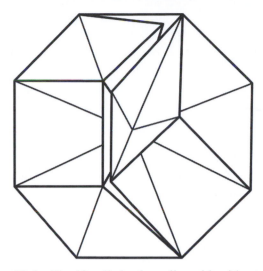

Figure 12.6   The 12-celled polar collapsoid, without flaps.

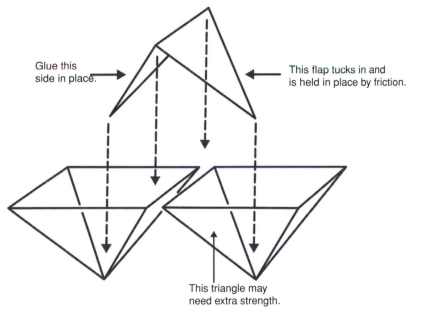

Glue this
side in place.

This flap tucks in and
is held in place by friction.

This triangle may
need extra strength.

Figure 12.7   Positioning a flap.

Practical hints

It may happen that, as you complete the model by sliding the flaps into place, you
observe that a triangular face of the cell into which you want to tuck the flap seems
a little flimsy. If so, glue another tab around this face. The result will be a very
sturdy cell into which you can now tuck the flap. This hint is useful for making any
of the collapsoids (see Figure 12.7).

(a)

(b)

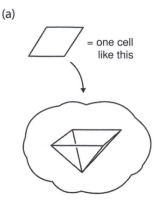

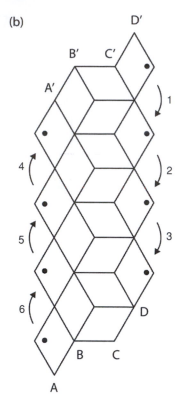

Figure 12.8   Diagram for constructing the 20-celled polar collapsoid.

12.4 Constructing a 20-celled polar collapsoid

As before, Figure 12.8 represents a net of baseless pyramids. Put the numbers 1 through 20 on the cells in the net and on your own cells. Begin this construction by joining 8 cells you have numbered in the net between B′C′ and BC in a string. These are the cells in the zone going from B′C′ to BC in the net diagram. Then continue by joining the cells on the right and left so that, as you look down on the figure, (1) you are looking into each cell, and (2) the outline of the cells looks precisely like Figure 12.8(b).

If you feel it would be helpful, label the triangles at the head and tail of each arrow with an identifying number (as shown in Figure 12.8(b)). Then use the tabs to join together the cells with like numbers. The vertices of the cells labeled with heavy dots will then be next to each other. You will notice that an edge going from one heavy dot to A(A′) to B(B′) to C(C′) to D(D′) and ending at the other heavy dot will remain open.

Collapse the model into $\frac{5}{6}$ of a regular hexagon and bring it back into expanded position several times until you understand the mechanics of its motion.

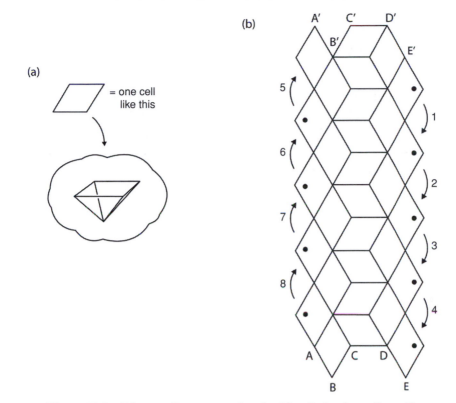

Figure 12.9   Diagram for constructing the 30-celled polar collapsoid.

Attach the flaps alternately to one or the other of the loose sides along the open edge. Think of the edges as labeled 1, 2, 3, 4 and 5 as you traverse from the North Pole to the South Pole; then attach flaps to the left-hand side on edges 1, 3, and 5 and attach flaps to the right-hand side on edges 2 and 4.

You may need to reinforce the triangular faces onto which the flaps fit in the cells, as described in the earlier practical hint.

12.5 Constructing a 30-celled polar collapsoid

Figure 12.9 represents a net of baseless pyramids. Again, you may find it helpful to number the cells in the net and your own cells with the numbers 1 through 30. Begin this construction by joining 10 cells together in a string. These are the cells that go around the model from C′D′ to CD. Then join the cells on the right and left of this zone so that as you look down on the figure you are looking into the cells, and the outline of the cells looks precisely like Figure 12.9(b).

Next, the sides of the cells at the head and tail of the arrows should be joined to each other (so that the vertices bearing a heavy dot come together). The edge

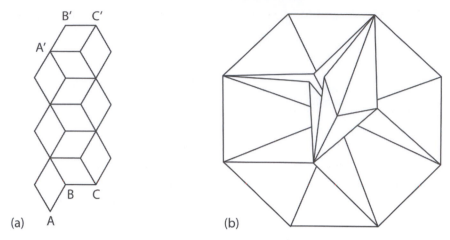

Figure 12.10   (a) The net diagram. (b) The finished collapsoid as viewed from one of its poles.

going from one heavy dot to A(A′) to B(B′) to C(C′) to D(D′) to E(E′) and ending at the other heavy dot will remain open. This model collapses into the shape of a complete regular hexagon.

Flaps may be attached to the open edge in the same alternating fashion as described for the 12- and 20-celled collapsoids. That is, think of the open edges as though they were labeled 1, 2, 3, 4, 5, and 6 as you traverse from North Pole to South Pole on this model; then attach flaps to the left-hand sides on edges 1, 3, and 5 and to the right-hand sides on edges 2, 4, and 6. As before, it may be necessary to reinforce a triangular face of a cell before tucking in the flap.

12.6 Constructing a 12-celled equatorial collapsoid

Figure 12.10(a) represents a net of baseless pyramids. Each parallelogram in the net represents one cell. Begin the construction by joining 6 cells together in a string. These are the cells that go along one zone (the equator of this model) between B′C′ and BC.

Next join the cells on the right and left so that as you look down on the figure you are now looking into the cells, and the outline of the cells looks precisely like the net in Figure 12.10(a).

Now join the side B′C′ to BC and then the side A′B′ to AB. You now have a 12-celled collapsoid, as shown in Figure 12.10(b), that will fold flat about the ring of 6 cells forming the equatorial zone.

Flaps added to the 3 cells on either side of the equatorial zone will help to keep the model in its inflated form. Make the flaps all go in either a clockwise or counterclockwise direction about each pole. And, as with the other models, you may wish to reinforce the triangle onto which the flap falls when it is in its final position.

Challenges

Now that you have constructed your polyhedra, you should get to know them. If you color them in various ways you will learn a great deal about their symmetries and how they are related to other polyhedra in this book.

One way to color the models is to get some gummed colored paper, available at art stores or office supply stores, and prepare a number of 2-triangle tabs in assorted colors. These may then be glued on top of the faces you wish to color. If gummed paper is not available, ordinary colored paper may be glued onto the faces.

When we talk of coloring an edge, we mean gluing a colored tab over that edge. The effect of this gluing will be that 2 adjacent triangles on the surface of the collapsoids are colored.

We now give you some specific suggestions for coloring.

For any collapsoid

Color the zones

Color one convex edge[2] red, for example; then color the edge opposite that edge (in the same cell) red, and the edge opposite that edge red, . . . until you have colored all of the edges in that zone. Then begin again on any uncolored edge and repeat the process with another color, and so on, until all edges are colored (and, hence, all faces!). You will then be able to see very clearly that the 12-, 20-, and 30-celled collapsoids have 4, 5, and 6 zones, respectively.

For the 12-celled collapsoid (polar or equatorial)

Color the cube

About a vertex where 4 convex edges come together, color each of those edges red, for example, and also color red the 4 convex edges surrounding the diametrically opposite vertex. Then begin again at any other uncolored vertex surrounded by 4 convex edges and color those edges blue, for example, and also color blue the 4 convex edges surrounding the diametrically opposite vertex. There will remain just 2 diametrically opposite pairs of vertices surrounded by 4 uncolored convex edges. Color those 8 edges with a third color. Compare this model with a cube!

---

[2] By a convex edge, we mean an edge that would be in contact, along its entire length, with a tight elastic material sheathing the collapsoid. Alternatively, we call an edge of a collapsoid *convex* if we can rest the collapsoid on a flat surface with that edge touching the surface.

Color the octahedron

About a vertex where 3 convex edges come together, color each of those edges red, for example, and also color red the 3 convex edges surrounding the diametrically opposite vertex. Then begin again and repeat the process with another color. Do this two more times. Compare this model with an octahedron!

For the 30-celled collapsoid

Color the dodecahedron

About a vertex where 5 convex edges come together, color each of those edges red, for example; then color red the 5 edges surrounding the diametrically opposite vertex. Repeat this process, using a new color each time, until you have colored all 60 convex edges. Compare this model with a dodecahedron!

Color the icosahedron

About a vertex where 3 convex edges come together, color each of those edges red, for example; then color red the 3 edges surrounding the diametrically opposite vertex. Repeat this process, using a new color each time (you will need 10 colors), until you have colored all 60 convex edges. Compare this model with an icosahedron!

12.7 Other collapsoids (for the experts)

Both the 20- and the 30-celled collapsoids can be made in an equatorial form – but not with equilateral triangles. To see this, notice that the 20-celled collapsoid has 8 cells in each zone, and the 30-celled collapsoid has 10 cells in each zone. Thus the 20-celled equatorial collapsoid must be made from cells that are parts of an *octagon*, and the 30-celled equatorial collapsoid must be made from cells that are parts of a *decagon*. In each case, when you construct the model with the appropriate cells, you proceed as before. The only difference is that, just as in the case of the 12-celled equatorial collapsoid, you ignore the arrows, and, instead, connect the ends of the principal zone (the one you first constructed). In this way you get a flower-like arrangement about both poles, which may be held shut with paper clips; the entire model will collapse about the equatorial zone. Since you know how to fold both regular 8- and 10- gons, we may confidently leave the exploration of these models to our very enthusiastic readers!

12.8 How do we find other collapsoids?

First we need to prove that collapsoids must have $n(n-1)$ cells. The argument comes from Coxeter [8] concerning the number of faces on a zonohedron. His argument goes as follows:

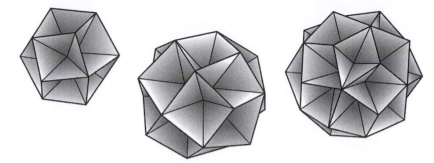

(a) 12-, 20- and 30-celled collapsoids.

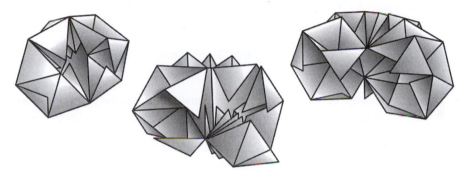

(b) Partly collapsed.

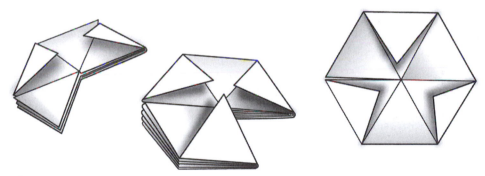

(c) Collapsed.

Figure 12.11   Collapsoids in various states.

These rhombic figures suggest the general concept of a *convex polyhedron bounded by parallelograms*. We proceed to prove that such a polyhedron has $n(n-1)$ faces, where n is the number of different directions in which edges occur.

Since all the faces are parallelograms, every edge determines a *zone* of faces, in which each face has two sides equal and parallel to the given edge. Every face belongs to two zones which cross each other at that face and again elsewhere (at the 'counter-face'). Hence the faces occur in opposite pairs which are congruent and similarly situated in parallel planes.

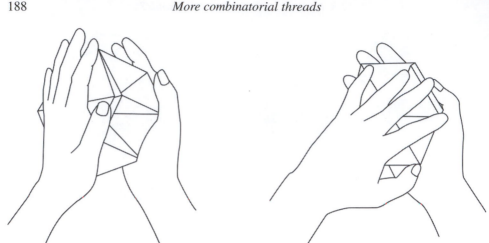

Figure 12.12   Collapsing the 12-celled polar collapsoid.

So also, the edges occur in opposite pairs which are equal and parallel and the vertices occur in opposite pairs whose joins all have the same mid-point. In other words, the polyhedron has *central symmetry*. Hence each zone crosses every other zone twice. If edges occur in n different directions, there are n zones, each containing $n - 1$ pairs of opposite faces, and hence $\binom{n}{2}$ pairs of opposite faces altogether. In fact, for every two of the n directions there is a pair of faces whose sides occur in those directions. Thus there are $n(n - 1)$ faces.

Since the collapsoids are constructed by replacing the parallelogram faces by cells it follows that every collapsoid must have $n(n - 1)$ cells. We have already discussed the cases where $n = 4, 5, 6$. What would be nice is to be able easily to find others.

We proceed by looking at the relationship between the 12-celled collapsoid and the cube (or octahedron). Observe that a skew equal-sided quadrilateral can be drawn across each edge of a cube such that two opposite vertices are coincident with adjacent vertices of the cube and the other two vertices are midpoints of adjacent faces on the cube. One such quadrilateral is shown with broken lines in Figure 12.13(a). Now suppose all of these skew quadrilaterals to be replaced with the cell made from equilateral triangles. What results is the 12-celled collapsoid, which has 4 zones.

Likewise, suppose you followed the same procedure, drawing skew equal-sided quadrilaterals across each edge on the octahedron, as shown in Figure 12.13(b), and then replaced each quadrilateral with a cell made from equilateral triangles. You would also get the 12-celled collapsoid, which has 4 zones.

Proceeding in precisely the same way, beginning with either the dodecahedron or the icosahedron (both of which have 30 edges), you can see how to design a 30-celled collapsoid, having 6 zones. See Figure 12.14.

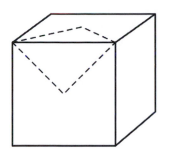
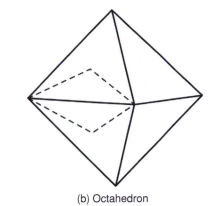

(a) Cube   (b) Octahedron

Figure 12.13 Designing a 12-celled collapsoid.

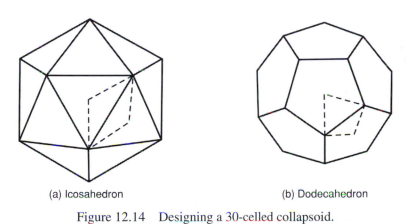

(a) Icosahedron   (b) Dodecahedron

Figure 12.14 Designing a 30-celled collapsoid.

The above procedure suggests the following question:

For how many other solids, of any kind, can the surface be sectored this way, that is, into $n(n-1)$ skew quadrilaterals having one diagonal on the edge of the given polyhedron, such that you can replace those quadrilaterals with baseless pyramids (cells) and get a collapsoid?

To answer this question we begin by making the observation that if you take the 30-celled collapsoid and remove one zone, it turns out that you get the 20-celled collapsoid, that has 5 zones. So, what we look for is a polyhedron whose number of edges is of the form $n(n-1)$ with n as large as possible.

A little checking among the semiregular polyhedra, known as the Archimedean solids, shows that the truncated icosahedron has 90 edges. If each edge is replaced with a cell, using the method above, a pseudo-enneacontahedron, or 90-celled collapsoid will result. The net for this 10-zoned collapsoid is shown in Figure 12.15.

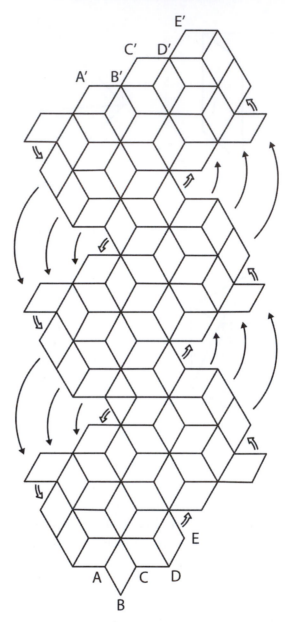

Figure 12.15   Net for the pseudo-enneacontahedron, a 90-celled collapsoid.

The geometry of any collapsoid dictates that you only need to make the cells for its net from parts of *appropriate* regular polygons if you want your models to fold up about the polar axis, or around an equatorial zone. If this is not a consideration, the cell made from equilateral triangles can always be used and produces nice, but not easily collapsible, models when *n* is greater than 6. This is because, when the

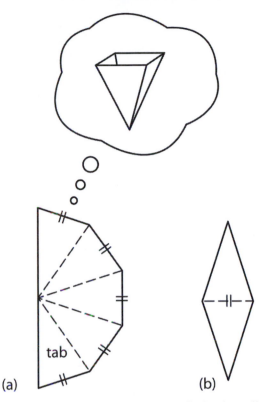

Figure 12.16 Pattern pieces for the 90-celled polar collapsoid.

number of edges between opposite vertices on a collapsoid exceeds 6, the cells must spiral on themselves in order to fold up – and the thickness of the paper causes trouble. We leave to the reader the calculation as to what kinds of cells are required for the various collapsoids, by taking into account the geometry involved. For more details on this consult [57]. We close this chapter by showing how to construct a 90-celled collapsoid, and then how to use that net to get $n(n-1)$-celled collapsoids for $n = 9, 8, 7, 6, 5$, and 4 – in that order.

If you want a collapsoid with n zones (and hence $n(n-1)$ cells) to fold up about a polar axis, you must make the cells out of parts of a regular polygon with at least n edges, because this collapsoid has n edges between every 2 diametrically opposed vertices and those n edges must fold flat around the circumference of the figure in its collapsed state. On the other hand, you will see that an n-zoned collapsoid has an equatorial zone of $2(n-1)$ cells. Thus, for equatorial collapsoids, the regular polygon from which the cells are constructed must have exactly $2(n-1)$ edges.

The 90-celled collapsoid which collapses about the polar axis is a magnificent model. When each of the cells is made from the pattern piece shown in Figure 12.16 it collapses into a complete decagon.

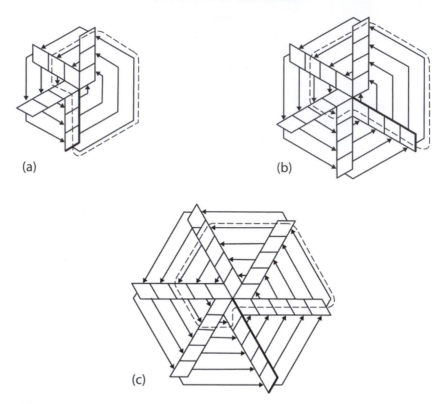

(a)

(b)

(c)

Figure 12.17   The heavy line indicates the open edge for polar collapsoids.

Now note that if you constructed the 90-celled collapsoid from the net in Figure 12.15, using cells made from equilateral triangles, and then removed one zone at a time, you would get a 72-, 56-, 42-, 30-, 20-, and finally, a 12-celled collapsoid. This would enable you to find nets for the ones not discussed here. You would also be able to note that the 72-, 56-, and 42-celled collapsoids are not as regular in appearance as the 12-, 30-, or 90-celled collapsoids. If you pursue this procedure more than once, you can see that not all of the collapsoids produced along the way are unique! There is, for example, a 30-celled collapsoid having the net shown in Figure 12.17(c). This polyhedron is different from the one produced by the net in Figure 12.9. Figures 12.17(a) and (b) show alternative representations for the nets of the 12- and 20-celled collapsoids, respectively. However, collapsoids produced by these nets are the same as those previously given for those models. Nevertheless, they are useful, because if we examine the three nets of Figure 12.17 they suggest than an *n*-zoned collapsoid can always be constructed by joining *n* strings of $(n - 1)$ cells each about a point. Indeed, this is the case, and we thus get an infinite class of collapsoids. This type of collapsoid is pineapple-like in appearance.

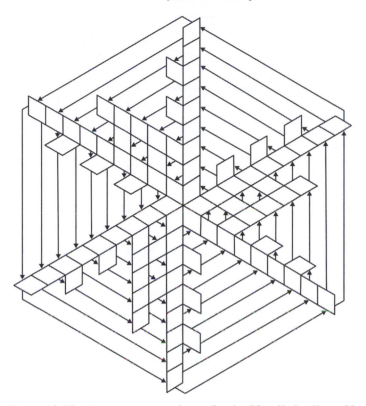

Figure 12.18   A more symmetric net for the 90-celled collapsoid.

The nets in Figure 12.17 are especially useful in identifying the location of each zone *before* you construct the model. The dotted lines show how this is done for one zone in each of the three nets. Colored pencils can be used to trace each zone on any given net; then the model can be assembled directly with the appropriately colored tabs.

Notice also that if the cells in the zone identified by the dotted line in Figure 12.17(c) were removed from the net and the remaining cells joined so as to preserve the other zones, the result would be the net shown in Figure 12.17(b). In a similar manner the net of Figure 12.17(b) produces that in Figure 12.17(a) when the identified zone is removed.

The net shown in Figure 12.18 is another representation of the net which produces the 90-celled collapsoid. It produces exactly the same polyhedron as the net shown in Figure 12.15. Apart from being prettier, it is also somewhat easier to trace zones on it. In fact, it is a convenient figure to use for finding the nets of 72-, 56-, and 42-celled collapsoids without actually having to construct them. Just identify each zone with its own color, remove one of the zones and redraw the remaining cells, joined so as to preserve the other zones.

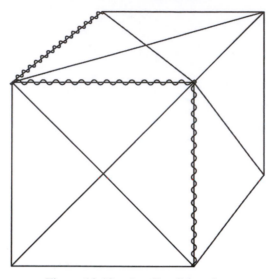

Figure 12.19   A collapsible cube.

The existence of 2 distinct 30-celled collapsoids is now evident. There are, in fact, at least 4 distinct 30-celled collapsoids, though no other versions of the 12- or 20-celled collapsoids exist. We leave the investigation of these ideas, and a general question, for the reader to ponder: ***How many different $n(n-1)$ collapsoids are possible for $n = 6, 7, \ldots$?***

We end this chapter by describing some very easily constructed collapsoids of a somewhat different type. You may have realized that the ordinary cube is a zonohedron. So it is natural to ask: *Will the cube collapse if we make it like other collapsoids?* The answer is both yes and no. It is impossible to replace each face with one of our cells made from equilateral triangles (why?), so in this respect we get a negative answer. However, it is possible to construct the cube from a special net on which every face has been folded along the diagonal lines before we assemble it. If such a cube is constructed and then cut apart along the line indicated in Figure 12.19, it *will* collapse in the expected manner. Try it!

Of course, this brings up another question: *Would the rest of the ordinary zonohedra collapse if they were made from faces that were scored along both diagonals and then cut apart along a line following an edge from one vertex to its diametrically opposite vertex?* The answer is yes!

13

Group theory – The faces of the trihexaflexagon[1]

13.1 Group theory and hexaflexagons

We described how to build a variety of hexaflexagons in Chapter 1. Here we give another description of how to build the 3-faced flexagon which has designs on each of its faces that enable us to track the set of motions bringing the flexagon into coincidence with itself.

The particular hexaflexagon we will consider in this chapter is the trihexa-flexagon (also called the 3-6-flexagon),[2] so named because it has 3 *faces*; that is, in any given state of the flexagon, one face (consisting of 6 equilateral triangles) will be up, one face will be down and one face will be hidden. Although the orientation of the faces will vary from state to state, the same 6 triangles will always appear together on a face.

By drawing a human visage on each face of the flexagon, and using a different color for each face, we can keep track of all the possible positions of the flexagon as it lies in a plane. We are thereby able to discover that the set of motions of this flexagon which bring it into coincidence with itself constitutes the dihedral group D_{18}.

13.2 How to build the special trihexaflexagon

The trihexaflexagon is constructed from a strip of paper containing 10 equilateral triangles[3] as shown in Figure 13.1. In order that the final model will flex easily the fold lines between the triangles should be creased firmly in *both* directions.

---

[1] The content of this chapter first appeared in an aritlce by Hilton, Pedersen, and Walser [48] and Pedersen wrote about the subgroup S_3 in [56]. We use the original illustrations with permission from the MAA.

[2] We may refer to the trihexaflexagon as simply the "flexagon" if no confusion would result.

[3] Notice that, from the point of view of decorating this piece, we have available a total of 20 triangles (because the strip of paper has 2 surfaces, the top surface and the bottom surface). When 2 of the triangles are glued to each other there remain 18 triangles with which to form the 3 faces.

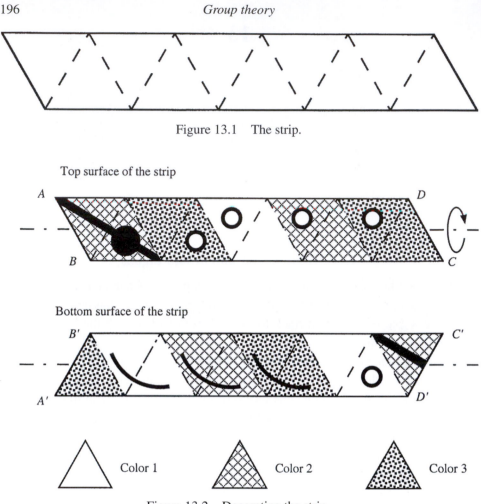

Figure 13.1   The strip.

Figure 13.2   Decorating the strip.

Now we decorate the strip as shown in Figure 13.2, where we make the bottom surface of the strip visible by flipping the entire pattern piece over a *horizontal* axis as indicated by the figure (where the vertices A, B, C, D should correspond with A′, B′, C′, D′, respectively, after you have flipped the piece over). ***Caution:*** Be careful here! Flipping the pattern piece over a *vertical* axis, and then decorating it as shown does ***not*** produce the desired flexagon.

Now we suggest that you view the construction of the flexagon as a puzzle. Here are some hints for constructing the flexagon with smiling (and frowning) faces.

(1) The first triangle on the upper portion of the strip is ultimately glued to the last triangle on the bottom portion (and it doesn't matter which one is on top of which). We suggest that you attach these triangles with a paper clip at first, and save the actual gluing until you are certain about the correctness of the construction.

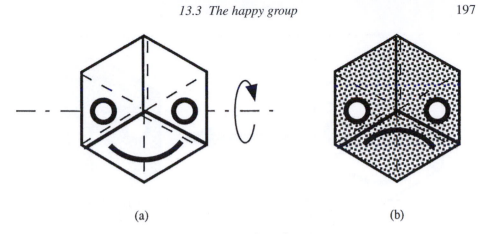

(a)   (b)

Figure 13.3   The trihexaflexagon.

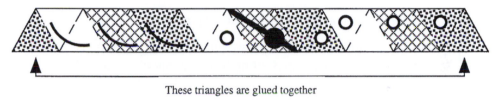

These triangles are glued together

Figure 13.4   The entire surface of the strip.

(2) The completed flexagon should show the visage of a smiling face, all of color 1 as you lay it down as shown in Figure 13.3(a); and when you flip the flexagon over, about a horizontal axis, it should show the visage of a frowning face, all of color 3 oriented as displayed in Figure 13.3(b).

(3) The strip that created the hexagon contains three half-twists; thus, like the Möbius band, it has only one surface (or side). Geometrically, this means there will be 3 slits on any face of this flexagon, symmetrically located at 120° intervals about its center. These slits are created by edges of the strip that go from alternate vertices of the hexagon to its center as shown in both parts of Figure 13.3.

From the last hint above we know that the flexagon now has only one surface. After you become proficient at manipulating your flexagon you may wish to verify with your own model that the repetitive pattern of 3 mouths, 3 right eyes, and 3 left eyes in the colors 1, 2, 3, respectively, occurs as shown in Figure 13.4.

13.3 The happy group

First we will always need to start with the flexagon in a ***standard initial position***, that is, with the smiling face of color 1 up and oriented precisely as shown in Figure 13.5(a).

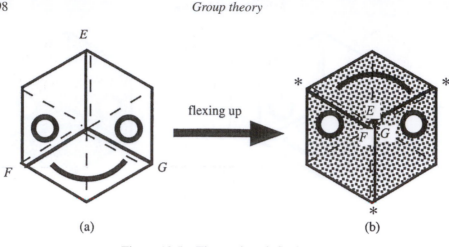

Figure 13.5   The motion f, flexing up.

Now we assume $n \geq 0$ and define the following motions:

$$\begin{cases} \text{the identity motion 1, which means we retain the initial position,} \\ f = \text{the motion of flexing up, starting from the initial position,} \\ f^n = \text{the motion of flexing up } n \text{ times, starting from the initial position.} \end{cases}$$

More precisely, the motion f consists of lifting vertices of the hexagon labelled E, F and G in Figure 13.5(a) above the flexagon until they meet, when the flexagon will come apart at the bottom and fall into the shape of a new hexagon with the vertices E, F and G at its center. If this is done correctly (it is important not to rotate the flexagon in either direction), you will see the upside-down smiling face of color 3 shown in Figure 13.5(b). Notice that the slits in the flexagon have revolved $\frac{1}{6}$ of a turn. Thus, when you flex up the second time you will have to bring the vertices marked with asterisks (*) together above the flexagon. A simple way to remember what to do is that, in each case, the vertex at the forehead of the human visage gets lifted to the center (and it disappears as the motion is completed).

We now follow the usual, obvious procedure of identifying a motion with its effect on the initial position. When we do this we see that $f^{18} = 1$, the identity motion.

Once you have mastered the motions f^n, you may verify the sequence of motions which produce the happy group shown in Figure 13.6; here we have adopted the identification indicated above.

Since f^{18} is the identity, we see that the happy group is the cyclic group C_{18}, generated by f.

Next we define *flexing down*. To describe this motion $\overline{f}$, we begin, as before, with the flexagon in the standard initial position shown[4] in Figure 13.7(a). Then $\overline{f}$

---

[4] This is, of course, the same initial position as that in Figure 13.5(a).

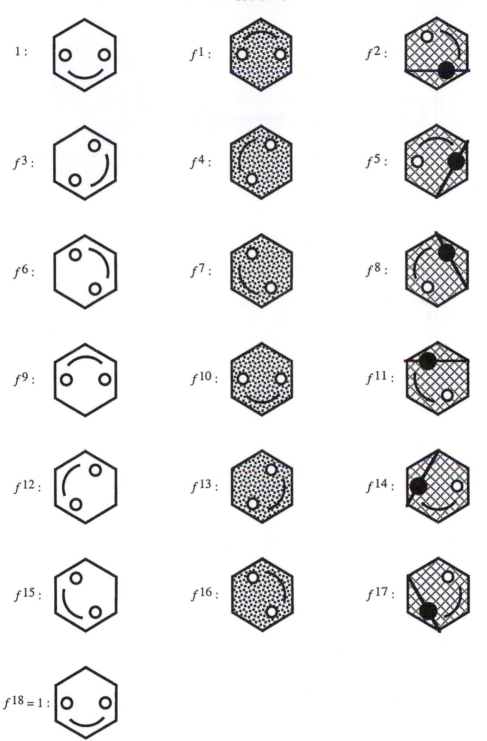

Figure 13.6   The happy group.

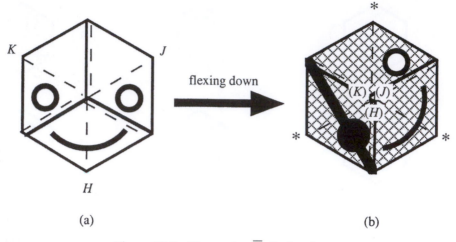

Figure 13.7   The motion $\overline{f}$, flexing down.

means that we push the vertices of the hexagon labeled H, J and K downwards until they meet; at that stage the flexagon will come apart at the *top* and fall into the shape of a new hexagon with the vertices H, J, and K at its center, but underneath the hexagon (this is indicated by the fact that these letters H, J, K appear in parentheses). If this is done correctly, we will obtain the smiling pirate face of color 2 as shown in Figure 13.7(b). Just as with the up-motions, it is important not to rotate the flexagon in either direction as we flex it. To obtain $\overline{f}^{n}$, we simply repeat the process of flexing down n times (notice that when we flex down the second time it is the vertices labeled with the asterisk (*) that come together beneath the flexagon). It is interesting that, in flexing down, the vertex at the forehead of the human visage moves up (as when flexing up), but in this case the flexagon visibly splits across the forehead before it falls flat, revealing the pirate.

Beginning with the flexagon in the standard starting position, you may verify that $\overline{f}$ yields the same face as f^{17} of Figure 13.6. Thus $\overline{f} = f^{17} = f^{-1}$. This means that, if you start with the position indicated on the right of Figure 13.7 and flex down, you get the initial position. In other words, flexing up is the inverse of flexing down (and vice versa), as you might expect.

If you're enjoying this you may check your flexing skill by reversing all of the steps of the happy group in Figure 13.6!

13.4 The entire group

We realize that the full group for this flexagon must be larger than C_{18} because no frowning faces ever appeared under the motions f^{n}. Cheerful as this situation is, it is plainly not complete. Like everything in this world this flexagon has good

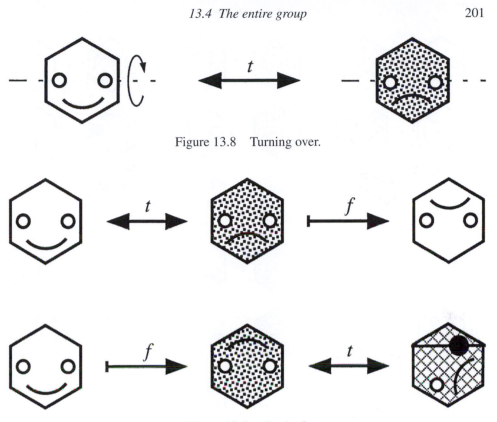

Figure 13.8 Turning over.

Figure 13.9 $ft \neq tf$.

(happy) and bad (unhappy) features. In order to get the entire group we certainly need to have a motion that makes the unhappy faces visible. To achieve this we introduce a new motion,

$$t = \text{turn over (so the rotation is about a } horizontal \text{ axis).}$$

Thus, if we begin with the flexagon in the standard initial position and perform the motion t we will see a frowning face of color 3 (see Figure 13.8).

Obviously t is an involution, that is, $t^2 = 1$. Figure 13.9 shows that the motion ft (meaning first do t, then do f) is not the same as tf (meaning first do f, then do t). Check this (remembering that the flexagon should be in the standard initial position, in both cases, when you apply the motions). Thus we see that our new motion t does *not* commute with f. We also notice that when the pirate frowns his patch covers his left eye, instead of the right one![5]

---

[5] Although we could do without any eye patches in analyzing this particular flexagon, it is clear that this feature may provide a better way of keeping track of the faces on more complicated flexagons – and we thought the pirate made an interesting addition to this group (both visually and socially!).

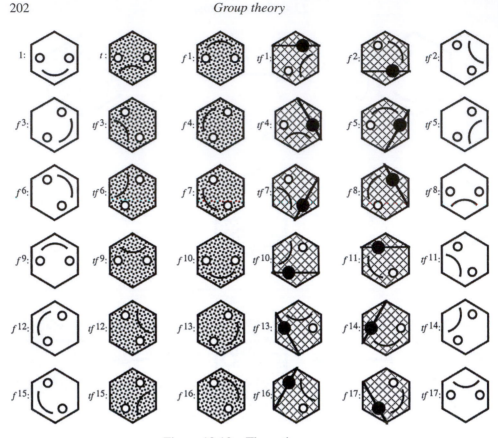

Figure 13.10   The entire group.

Figure 13.10 shows all the possibilities for f^n and tf^n. Notice that the first, third and fifth columns are just the smiling faces from Figure 13.6. This observation may give you an idea of an easy way to confirm that the visages in Figure 13.10 are correct.

We have already seen that $tf^n \neq f^n t$. However, since flexing *up*, as viewed from *above* the flexagon, is the same as flexing *down*, as viewed from *below* the flexagon we have,

$$f^n t = tf^{-n}.$$

Thus we see that the group generated by f and t has 36 elements and is therefore the full group of motions of our flexagon. Since the generators f, t satisfy the defining set of relations $f^{18} = 1, t^2 = 1, ft = tf^{-1}$, the group is the dihedral group D_{18}, the group of symmetries of the regular 18-gon (shown in Figure 13.11). Figure 13.12 shows the effects of the group elements where the single-headed arrows denote the f action and the double-headed arrows denote the t action. (This

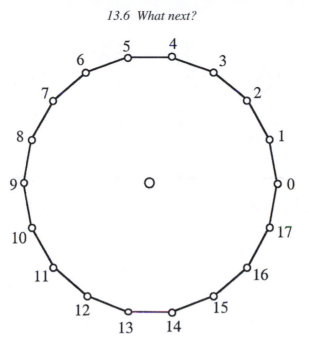

Figure 13.11 The regular 18-gon.

figure was, in fact, on the cover of the issue of *Mathematics Magazine* in which the original article [48] on this topic appeared.)

13.5 A normal subgroup

If we are only interested in the different expressions on the faces of our flexagon, without respect to orientation, we have only 6 cases (as seen in Figure 13.13), instead of 36.

Figure 13.10 motivates the following argument.

We obtain the group of motions of the unoriented faces by adding the relation $f^3 = 1$ to our group D_{18}. The resulting quotient group of D_{18} by the normal subgroup generated by f^3 is then generated itself by F and t, subject to $F^3 = 1$, $t^2 = 1$, $Ft = tF^{-1}$. Here, F is, of course, the image in the quotient group of f; and the quotient group is just the symmetric group S_3.

13.6 What next?

In [56] the trihexaflexagon was discussed and the group S_3 was obtained by using a flexagon where each of the 3 faces were simply different colors. In [18] the group D_9 was obtained by a systematic labeling of the vertices of the 6 triangles on each of the 3 faces of the trihexaflexagon. However, in order to obtain the entire group D_{18},

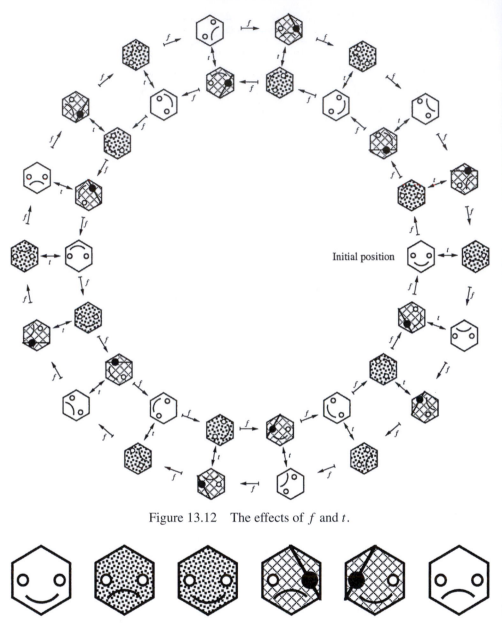

Figure 13.12   The effects of f and t.

Figure 13.13   Facial expressions of the flexagon.

it was necessary to introduce a finer method of distinguishing between the different orientations of the faces; distinguishing between smiling and frowning visages did the trick. The obvious next question to explore is whether this, or some refinement of it, will help to identify the mathematical structure of the hexahexaflexagon (with 6 faces).

It turns out that the trihexaflexagon is very special. According to Berkove and Dumont [4] they have "determined that the tri-hexaflexagon is exceptional, as it is the *only* member of the hexaflexagon family whose collection of motions forms a group." Berkove and Dumont then shifted their attention to square flexagons (that they called tetraflexagons) which are constructed with strips of squares folded into a 2×2 square final form. In their article they present some new results and open questions.

14

Combinatorial and group-theoretical threads – Extended face planes of the Platonic solids[1]

14.1 The question

The main purpose of this chapter is to ask, and answer by elementary methods, the question: *How many bounded and unbounded regions in space result when the planes of the Platonic solids are extended in space?* To set the scene we first discuss, in Section 14.2, the analogous question in the plane; that is, we ask, and answer, the question "how many bounded and unbounded planar regions result when the sides of regular polygons are extended in the plane." Then, in Section 14.3, we recall some of the properties of the Platonic solids that were discussed in Section 5.3.

In Section 14.4 we answer the main question, except in the case of the icosahedron, where we refer the reader to [50] or [76] in order to find out how others calculated the number of bounded regions that result when the face planes of the icosahedron are extended in space.

In Section 14.5 we suggest some questions for the interested reader to ponder. We wish to emphasize that there are no new results in this chapter (see [76]), but our approach is more elementary than the classical discussions.

14.2 Divisions of the plane

The reader might like to try answering the following question before reading further.

> **How many bounded and unbounded regions result when the sides of a regular *n*-gon are extended in the plane?**

[1] This chapter covers the material one of the authors (JP) talked about in her BAMA (Bay Area Mathematical Adventures) presentation at San Jose State University on February 6, 2002; an account of the talk was published in [63]. However, JP had previously published some of the content in [59]. JP is grateful to John E. Wetzel for bringing this problem to her attention (see [50]) and to Alexanderson and Wetzel [1], Rouse Ball [2], Coxeter [8], Cundy [11], Cundy and Rollett [12], Pólya [67], and Wenninger [79 and 80] for providing the stimulating material in their publications that enabled her to make the connection between the braided models and the main question of this chapter.

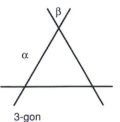

3-gon
1 bounded region
6 unbounded regions

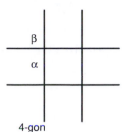

4-gon
1 bounded region
8 unbounded regions

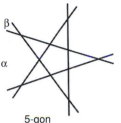

5-gon
6 bounded regions
10 unbounded regions

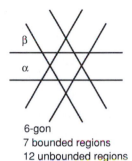

6-gon
7 bounded regions
12 unbounded regions

Figure 14.1 Regions created by straight lines in the plane.

Solution An examination of Figure 14.1 should convince you that the number of unbounded regions created by the extended sides of a regular n-gon, $n \geq 3$, is $2n$. You might also observe that, for any given n, the unbounded regions come in precisely two types (labeled as α and β in Figure 14.1) which alternate as you scan around the periphery of the figure.

The trickier part of the problem is to find the number of bounded regions. The problem seems to naturally break into the following two cases.

Case 1

Notice that when n is odd (with $n \geq 3$), each of the n sides, when extended, will intersect each of the other $(n-1)$ sides. Thus, when n is odd, think of drawing the extended sides that produce the final polygon, one at a time, and record the number of *new* bounded regions created at that stage. We then have Table 14.1 where, starting with the third line, the number of new finite regions increases one at a time until, finally, there are $(n-2)$ new regions created by the nth line.[2]

---

[2] It may help you to look at a particular but not special case, like the 7-gon, to see what this table means.

Table 14.1.

The introduction of the extended side number	creates the following *new* bounded regions
1	0
2	0
3	1
4	2
5	3
6	4
•	•
•	•
•	•
n	$n - 2$

Thus, summing all the entries in the second column, we see that the total number of bounded regions, B_n, for a regular n-gon, with n odd, is given by

$$B_n = 1 + 2 + 3 + \cdots + (n - 2). \tag{14.1}$$

Using the formula[3] $\sum_{i=1}^{N} i = \frac{N(N+1)}{2}$ with $N = (n - 2)$ we see that (14.1) can be rewritten rather neatly as

$$B_n = \frac{(n - 2)(n - 1)}{2}. \tag{14.2}$$

It is interesting to note that this formula is, in fact, true for $n \geq 1$, although, in some sense, the argument and formula (14.2) only started producing finite regions when $n = 3$.

Case 2

Now we turn to the case where n is even. We first observe that, in this case, each of the n sides when extended will intersect each of the other $(n - 2)$ sides that are not parallel to it. For convenience let us assume $n = 2k$, $k \geq 2$, and proceed much the same way as in Case 1, except that now we need to take into account the fact that, when we draw the $(k + 1)$st extended side, it will be parallel to the first extended side that was drawn. Thus we have Table 14.2, naturally broken into two parts, where, starting with the third line of Part 1 the number of new finite regions increases one at a time until, in the middle, there are $(k - 2)$ new regions created

---

[3] See [70] for a nice discussion of how to derive this and other summation formulas.

Table 14.2.

Part 1			Part 2		
The introduction of the extended side number		creates the following *new* bounded regions	The introduction of the extended side number		creates the following *new* bounded regions
1		0	$k+1$		$k-2$
2		0	$k+2$		$k-1$
3		1	$k+3$		k
4		2	•		•
5		3	•		•
•		•	•		•
•		•	$2k-2$		$2k-5$
•		•	$2k-1$		$2k-4$
k		$k-2$	$2k$		$2k-3$

by the extended kth side. Then in Part 2 – because the $(k+1)$st side is parallel to the first side – there are also $(k-2)$ new regions created by the extended $(k+1)$st side at the top of the second column in the table.

Thus, by summing all the entries in the second columns of Part 1 and Part 2, we see that the total number of bounded regions, B_{2k}, for a regular $2k$-gon is

$$B_{2k} = [1 + 2 + 3 + \cdots + (2k - 3)] + (k - 2), \qquad (14.3)$$

where the last term occurs because of the repeat of $k-2$ at the top of the table in Part 2.

Again we use the formula $\sum_{i=1}^{N} i = \frac{N(N+1)}{2}$, but this time with $N = 2k - 3$, to write (14.3) as

$$
\begin{aligned}
B_{2k} &= \frac{(2k-3)(2k-2)}{2} + (k-2) \\
&= (2k-3)(k-1) + (k-2) \\
&= (2k-2)(k-1) - (k-1) + (k-2) \\
&= 2(k-1)^2 - 1.
\end{aligned}
$$

Recalling that $n = 2k$, we see that B_n takes the following form:

$$B_n = \tfrac{1}{2}(n-2)^2 - 1, \quad \text{where } n \text{ is an } even \text{ number} \geq 4.$$

Table 14.3 *Some facts about the Platonic solids.*

p	q	Name of solid	V	E	F
3	3	Tetrahedron	4	6	4
4	3	Hexahedron (cube)	8	12	6
3	4	Octahedron	6	12	8
5	3	Dodecahedron	20	30	12
3	5	Icosahedron	12	30	20

14.3 Some facts about the Platonic solids

We recall, from Section 5.3, that there are precisely 5 types of regular convex polyhedra. These 5 polyhedra are known as the Platonic solids. The name of each polyhedron and the corresponding number of vertices (V), edges (E), and faces (F) are also shown in Table 14.3. The number of sides in each face of the polyhedron is p and the number of polygons that come together at each vertex of the polyhedron is q. Illustrations of the solids appear in Figure 14.2.

Notice that there is a striking dual relationship between the number of vertices (V) and the number of faces (F) – as well as the dual relationship between p and q. Important features of this duality are that the $\left({\text{hexahedron} \atop \text{dodecahedron}}\right)$ has the same number of faces as the $\left({\text{octahedron} \atop \text{icosahedron}}\right)$ has vertices; and the $\left({\text{octahedron} \atop \text{icosahedron}}\right)$ has the same number of faces as the $\left({\text{hexahedron} \atop \text{dodecahedron}}\right)$ has vertices. So we say that the hexahedron and the octahedron are ***duals*** of each other; similarly the dodecahedron and icosahedron are duals of each other. The tetrahedron is ***self-dual***, since it has the same number of vertices and faces ($V = F = 4$).

These models can be neatly distinguished from each other by what are called ***symmetry groups*** (see Chapter 8 of [24] for a discussion of symmetry groups), but it will be enough here for the reader to simply observe certain properties about the rotational axes for each of the models. By a ***rotational axis*** we mean an axis about which the model can be rotated a certain fraction of 2π and still occupy the same space as in its original position. With the models in hand it is not difficult to see that

(a) The tetrahedron has
    4 axes about which it can be rotated $\pm\frac{2\pi}{3}$
    and 3 axes about which it can be rotated $\frac{2\pi}{2}$.
(b) Both the hexahedron and the octahedron have
    3 axes about which they can be rotated $\pm\frac{2\pi}{4}$
    4 axes about which they can be rotated $\pm\frac{2\pi}{3}$

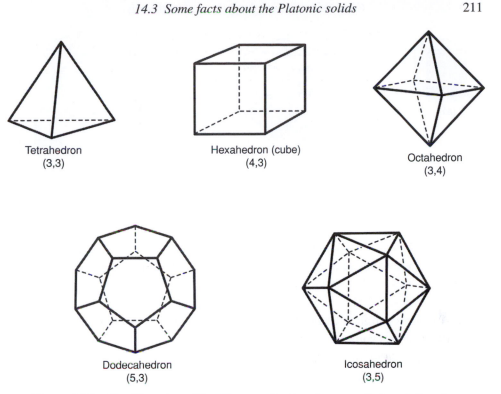

Figure 14.2   The Platonic solids: The notation (p, q) means that each face is a regular p-gon and q faces come together at each vertex.

and 9 axes (3 joining opposite faces of the cube or vertices of the
octahedron; and 6 joining the centers of opposite edges)
about which they can be rotated $\frac{2\pi}{2}$.

(c) Both the dodecahedron and the icosahedron have
6 axes about which they can be rotated $\pm\frac{2\pi}{5}$ or $\pm\frac{4\pi}{5}$
10 axes about which they can be rotated $\pm\frac{2\pi}{3}$
and 15 axes about which they can be rotated $\frac{2\pi}{2}$.

We say that the tetrahedron possesses ***tetrahedral symmetry***, the hexahedron and the octahedron possess ***octahedral symmetry***, and the dodecahedron and icosahedron possess ***icosahedral symmetry***. Models belonging to these three symmetry groups are easily distinguished from each other, since octahedral symmetry is the only one allowing rotations of $\frac{2\pi}{4}$, and icosahedral symmetry is the only one allowing rotations of $\frac{2\pi}{5}$.

There are, of course, several classical ways to construct these polyhedra see [2, 12, 80]. However, as we showed in Chapter 8, each of these models may be braided with straight strips of paper. The unexpected payoff is that these

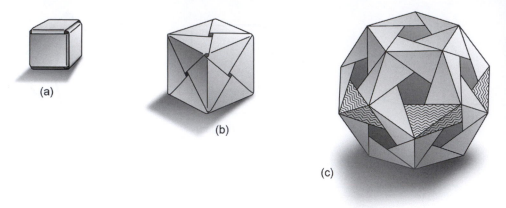

Figure 14.3 (a) A cube braided from 3 straight strips. (b) The diagonal cube braided from 4 straight strips. (c) The golden dodecahedron braided from 6 straight strips.

braided models can then be used to help answer the main question of this chapter. Figure 14.3 shows some of the braided models from Chapter 8 that will be playing an important role in the next section.

14.4 Answering the main question

We are now ready to turn to our main question: ***How many bounded and how many unbounded regions in space result when the planes of the Platonic solids are extended in space?***

Let us look at the solids in the order in which they appear in Table 14.3 above.

Tetrahedron

The tetrahedron is the only Platonic solid whose faces do not lie in parallel pairs of planes. We are able to answer the question for this polyhedron by brute force, using the illustration of the tetrahedron with its planes extended shown in Figure 14.4.

From Figure 14.4 we see that there are, in fact, 15 regions created by the extended face planes of the tetrahedron:

$$\begin{array}{ll}
& \text{1 bounded region, the tetrahedron itself with its interior} \\
+ \text{(a)} & \text{4 unbounded trihedral regions from its vertices} \\
+ \text{(b)} & \text{4 unbounded truncated trihedral regions from its faces} \\
+ \text{(c)} & \text{6 unbounded wedges from its edges} \\
= & \text{1 bounded region} + \text{14 unbounded regions.}
\end{array}$$

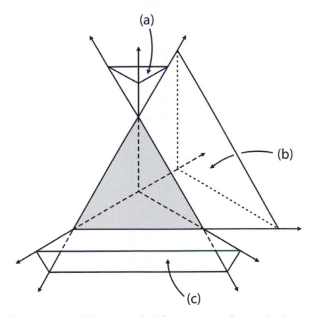

Figure 14.4   The extended face planes of a tetrahedron.

Hexahedron (or cube)

The extended face planes of the cube partition space into 27 pieces as seen in Figure 14.5:

$$
\begin{aligned}
& \text{1 bounded region, the cube itself with its interior}\\
&+ \text{(a)} \quad \text{8 unbounded trihedral regions from its vertices}\\
&+ \text{(b)} \quad \text{6 unbounded square prisms from its faces}\\
&+ \text{(c)} \quad \text{12 unbounded wedges from its edges}\\
&= \text{1 bounded region} + \text{26 unbounded regions.}
\end{aligned}
$$

But now we observe that *because* its faces lie in parallel pairs of planes we are able to make a braided model of it (Figure 14.6). Notice that the edges of the three strips used to create the braided model lie in 6 planes which intersect each other to form the surface of the cube. The details come later but, meanwhile, whenever braided models appear, please compare the various features (holes, 1-thickness regions, and 2-thickness regions) of the braided model with the same features of the associated unbounded regions.

Octahedron

The face planes of the regular octahedron, when extended, form a slightly more complicated division of space. Figure 14.7 shows how this division takes place.

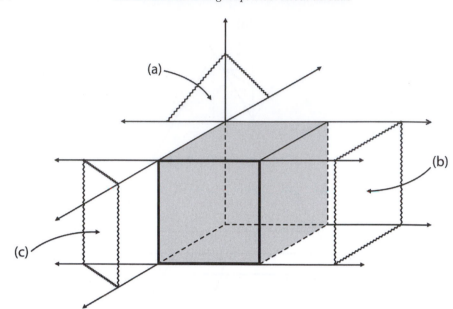

Figure 14.5   The extended face planes of the hexahedron (cube).

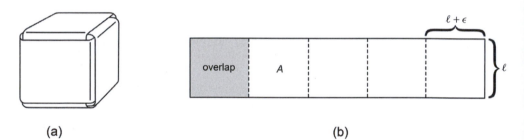

(a)   (b)

Figure 14.6   (a) A braided cube, with (b) one of the strips from which it is made. (ϵ represents a small quantity as compared to ℓ.)

First, 8 tetrahedra, like those marked (a) appear on the octahedron's faces; the visible surfaces of these 8 tetrahedra form the ***stella octangula***.[4] Then the unbounded regions are formed. The bubble in Figure 14.7 shows the stella octangula and the main figure shows two of the tetrahedra and one each of the various kinds of unbounded regions. Figure 14.8 shows how the unbounded regions are related to the various regions on the surface of the braided diagonal cube.

A new feature is that not all the unbounded regions grow straight out of a vertex, a face, or an edge of the original octahedron; some grow out of vertices, edges, or faces of one of the tetrahedra. The complete count is:

---

[4] This figure was noted by Johannes Kepler (1571–1630) about 1619 in his *Harmonia mundi*, Propositio XXVI.

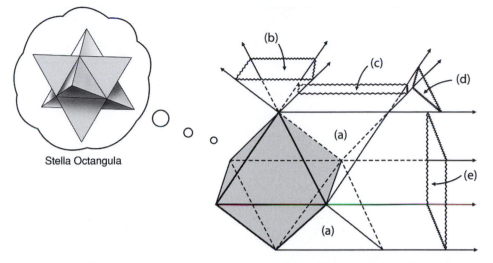

Figure 14.7 The extended face planes of the octahedron as related to the stella octangula.

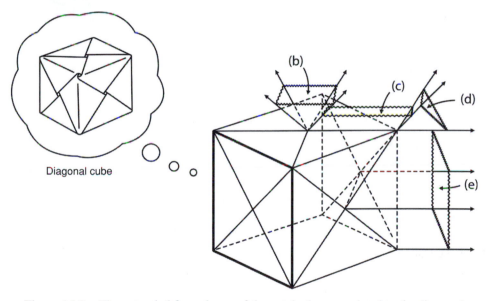

Figure 14.8 The extended face planes of the octahedron as related to the diagonal cube.

1 bounded region, the octahedron itself with its interior

+ (a) 8 bounded tetrahedral regions on its faces

+ (b) 6 unbounded regions with square cross-sections from its vertices

+ (c) 24 unbounded wedges from the edges with rectangular cross-sections of the 8 tetrahedra which do not coincide with edges of the original octahedron

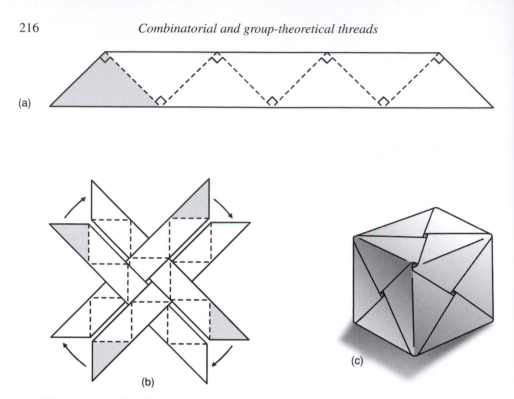

Figure 14.9   The diagonal cube showing (a) one of the 4 strips, (b) the layout for constructing the cube, and (c) the completed cube.

+ (d)   8 unbounded regions with triangular cross-sections from the outside vertices of the 8 tetrahedra

+ (e)   12 unbounded regions with quadrilateral cross-sections.

= 9 bounded regions + 50 unbounded regions.

If you have not already done so it may be helpful at this point to construct the diagonal cube shown in Figure 14.9 so that you may gain a better understanding of the geometry involved. The construction procedure is almost self-evident, especially if you remember that every strip must go alternately over and under the other strips on the model. It may help to secure the center square in the original layout with tape that can be removed once the model is finished. All the ends should tuck in neatly on the finished model.

But how does the octahedron fit in the diagonal cube? The clue lies in the planes formed by the "open" bases and tops of the 4 antiprisms made from 4 strips. These planes define 4 pairs of mutually intersecting parallel planes positioned symmetrically about a point. Since the braided model has octahedral symmetry, these 8 planes form the faces of a regular octahedron. Note that 4 planes, one from each pair, intersect at the center of each face of the diagonal cube; the 6 center points are the vertices of the original octahedron, and the 8 vertices of the cube are the outside vertices of the associated stella octangula shown in Figure 14.7.

Having the diagonal cube in hand enables you to read off from its surface the number and nature of the unbounded regions in space defined by the extended face planes of the octahedron. The same is true for the braided cube of Figure 14.6. Notice, for example, that the small square hole shown in the center of each face of the diagonal cube in the bubble of Figure 14.8 represents an unbounded polyhedral region, as does the small triangular hole at the vertex of the same diagonal cube. (The small triangular hole at the vertex of the cube in Figure 14.6(a) also represented an unbounded polyhedral region.) Notice, too, that where there is just one thickness of paper, along the diagonals of the faces of the diagonal cube, each of these regions represents an unbounded wedge. (The same was true for the single thickness that appeared along the edges of the cube in Figure 14.6(a).) Finally, notice that wherever there is a 2-thickness region on the braided model it represents an unbounded region with a polygonal cross-section. (A 2-thickness region appeared on each face of the cube in Figure 14.6(a).)

Now that we have seen the usefulness of the braided models in the cases of the cube and the octahedron, it should occur to us to take a careful look at the situations that can occur on the surface of any such model. This is, in fact, essential if one wishes to understand similar models for the dodecahedron and the icosahedron. Of course, since in this chapter we are only concerned with the Platonic solids, we impose the requirement that the individual strips of our braided models

(a) must have central symmetry;
(b) must cross over every other strip on the model at diametrically opposite places; and
(c) the entire braided configuration must have the symmetry group of the original surface being discussed.[5]

Because of restriction (b) no other lines of intersection between the planes defined by the edges of the strips can occur, and hence there can be no other bounded regions formed by extending the planes *outside* the braided model. Figure 14.10 then shows the three basic situations on the braided model that produce unbounded regions:

(i) An unbounded polyhedral region occurs whenever there is a hole in the braided model. The number of sides of the polyhedral region is easily determined by looking at the number of sides of the polygon surrounding the hole.

---

[5] In a more general setting, one could remove one of the strips in the diagonal cube, thereby reducing the symmetry group from that of octahedral symmetry to merely dihedral symmetry (see page 248 of [24] for details about this symmetry group) and use the resulting model to discover the unbounded regions of a rhombic hexahedron (think of a "squashed" cube).

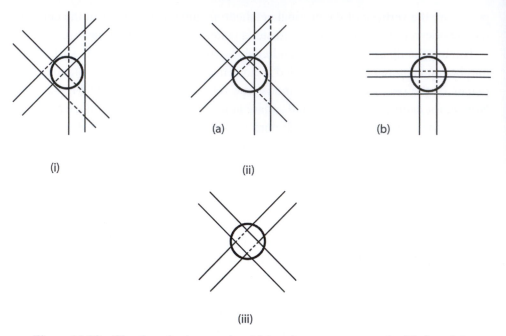

(i) (a) (ii) (b)

(iii)

Figure 14.10   The three basic ways in which strips can cross on a braided model.

(ii) An unbounded region with 4 unbounded faces and 4 unbounded edges; in the situation shown in (a) the unbounded regions will be "trapezium-like" while in the situation shown in (b) the unbounded regions will be "wedges off edges."
(iii) An unbounded "parallelogram-like" region with 4 unbounded faces and 4 unbounded edges occurs wherever there is a region of 2 thicknesses on the braided model.

Dodecahedron

Figure 14.11 at the top shows at its center the pentagonal face of the dodecahedron (Figure 14.11(a)). Next, at the top, is the star pentagram, 12 of which interpenetrate each other to form the small stellated dodecahedron of Figure 14.11(b). Then there appears a larger pentagon, 12 of which interpenetrate each other to form the great dodecahedron of Figure 14.11(c). Finally there is the larger pentagram, 12 of which interpenetrate each other to form the great stellated dodecahedron of Figure 14.11(d).[6] These stellations, in fact, produce all of the bounded regions created by extending the face planes of the dodecahedron. We can enumerate them as follows:

---

[6] In fact, the great stellated dodecahedron fits inside the golden dodecahedron with its outermost vertices touching the vertices of the golden dodecahedron (see Section 11.5 for details).

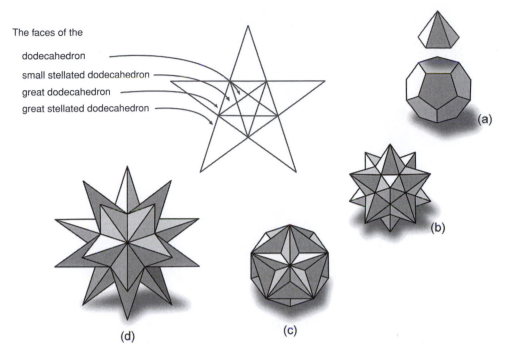

The faces of the

dodecahedron

small stellated dodecahedron

great dodecahedron

great stellated dodecahedron

(a)

(b)

(d)   (c)

Figure 14.11   Stellations of the dodecahedron.

from (a)   1 bounded region, the dodecahedron itself with its interior

from (b)   + 12 pentagonal pyramids on the dodecahedron's faces (producing the small stellated dodecahedron)

from (c)   + 30 wedge-like tetrahedra on the dodecahedron's edges (producing the great dodecahedron)

from (d)   + 20 triangular dipyramids on the dodecahedron's vertices (producing the great stellated dodecahedron)

            = 63 bounded regions.

We now use the golden dodecahedron of Figure 14.12 to find the unbounded regions. Notice that this model is formed of 6 strips. There are thus 6 pairs of parallel planes, and the zones between these pairs intersect in a "core" that has as its surface the regular dodecahedron.

Using the surface of the golden dodecahedron, we can count the unbounded regions formed by the face planes of the "core" dodecahedron as follows:

from (a)   12 unbounded pentahedral regions from the pentagonal holes on its faces

from (b)   + 20 unbounded trihedral regions from the triangular holes at its vertices

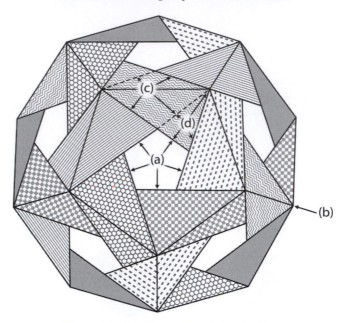

Figure 14.12   The golden dodecahedron.

from (c)   + 30 unbounded regions with "parallelogram-like" cross-sections from
             its edges

from (d)   + 60 unbounded regions with "trapezium-like" cross-sections from
             5 regions off the pentagonal holes on each of the 12 faces

           = 122 unbounded regions

Icosahedron

Not surprisingly, the situation for the icosahedron is the most difficult. The bounded
regions are very much more complicated, since, according to Coxeter, Duval,
Flather, and Petrie, there are 59 different kinds of intersections before the 20 face-
planes of the icosahedron go off into space (see [9]). According to an impressive
argument in [50] there are 473 bounded regions for the icosahedron. So let us take
this result as proved and try to see how to enumerate the unbounded regions for the
icosahedron.

Of course, what is needed is a braided model with 10 strips, having icosahedral symmetry. To see what motivated the discovery of this model the reader is
referred to another braided construction of the dodecahedron in Section 9.5 in
which the regular pentagonal dodecahedron is constructed from straight strips each
containing 6 trapezium sections. What is required, for producing a model to help
solve our current problem, is to make each of those 6 strips narrower so that 4

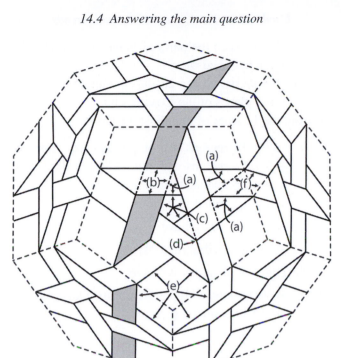

Figure 14.13   10 strips braided together about the ghost of a dodecahedron.

more can be introduced into the figure to complete the symmetry group for the docecahedron.

The model illustrated in Figure 14.13 can be used, without even constructing it, to obtain the information about the unbounded regions created by extending the face planes of the regular icosahedron. Why is this so? Well, first notice that the model surrounds the "ghost" of a regular dodecahedron. Furthermore, it has 10 strips since each strip crosses 6 of the ghost's faces, and each of these 12 faces must be crossed 5 times:

$$\frac{5 \times 12 \text{ (total number of crossings on the entire ghost)}}{6 \text{ (number of crossings contributed by each strip)}} = 10$$

(the number of strips on the model). For clarity, just one of the strips is shown shaded in Figure 14.13; of course, it crosses three other faces of the ghost on the back of the model.

Now notice that the edges of these 10 centrally symmetric strips identify 20 planes which must intersect inside the model to form some kind of centrally symmetric polyhedron. Finally, observe that the braided model has icosahedral symmetry. Thus we are forced to conclude that the polyhedron formed by the intersection of the zones between the pairs of planes determined by the edges of these 10 strips must be the icosahedron.

With Figure 14.13 before us we can easily enumerate the unbounded regions made by the extended face planes of the icosahedron.

First, coming from the 12 faces of the ghost we have:

from (a) 15×12 regions with trapezium-like cross-sections;
from (b) $+ 5 \times 12$ regions with parallelogram-like cross-sections;
from (c) $+ 1 \times 12$ pentahedral regions;
from (d) $+ 5 \times 12$ trihedral regions;

then from the vertices of the ghost we have

from (e) $+ 20$ hexahedral regions;

and, finally, from the edges of the ghost we have

from (f) $+ 30$ regions with parallelogram-like cross-sections
             $= 362$ unbounded regions.

14.5 More general questions

The models mentioned in this chapter are not the only braided models possible. For example, 4 identical straight strips can be braided to form a model which looks like a truncated octahedron (with holes instead of square faces) and the intersection of the planes determined by the edges of the braided strips is the regular octahedron. Likewise, 6 identical straight strips can be braided to form what looks like a truncated icosahedron (with holes instead of pentagonal faces) and the intersection of the planes determined by the intersection of those strips is a regular dodecahedron. You may find it challenging to try to discover *all* the braided models, for each of the Platonic solids, which have faces that lie in parallel planes.

For a real challenge you might wish to find analogous braided models for the remaining 13 *Archimedean solids* (the semiregular polyhedra whose edges are of the same length and whose faces are all regular, though not necessarily of the same shape; in addition the faces form the same arrangement about each vertex; see [2, 11, 12, 49, 80] for more details) – or prove that such a model cannot exist. As a hint note that not all of the Archimedean solids have faces that lie in parallel planes. The authors would be very interested in any new results you discover for these polyhedra.

15

A historical thread – Involving the Euler characteristic, Descartes' total angular defect, and Pólya's dream

15.1 Pólya's speculation

The authors were very privileged to have been on the friendliest of terms, during the later part of his long life, with the outstanding mathematician and teacher George Pólya – and with his equally remarkable wife Stella. Pólya would often muse out loud about the inspired work of the great mathematicians of the past and wonder how they got their brilliant ideas, and what was the source of their profound insights.

Once, in such speculative mood, Pólya was discussing the astonishing observation of Euler (see [14]) that, if one takes any convex polyhedron in the most simple, geometric-combinatorial sense, and counts the number of vertices V, the number of edges E, and the number of faces F, then

$$V - E + F = 2. \tag{15.1}$$

The nature of the formula (15.1) indicates that Euler was thinking of the polyhedra as 2-dimensional, and indeed, in modern terminology, as homeomorphic to the (2-dimensional) sphere S^2.

Pólya speculated that Euler would have been guided by his knowledge of the 1-dimensional case; in other words, of polygons (in the sense of closed polygonal paths) homeomorphic to the circle S^1. He imagined Euler saying to himself: We regard two polygons as equivalent if the vertices of one can be matched with the vertices of the other, and the edges of one with the edges of the other so that the incidence relations *vertex v belongs to edge e* are preserved. Then two polygons are equivalent if and only if they have the same number of vertices. Now it is easy to generalize the notion of equivalence – we regard two polyhedra as equivalent if the vertices, edges, and faces of one can be matched with the vertices, edges, and faces, respectively, of the other, so that the incidence relations *vertex v belongs to edge e*, *edge e is a side of face f* are preserved. How then are polyhedra to be classified?

Pólya dreamed that Euler might have tried to find a criterion of equivalence similar
to that which is valid for polygons. In his search, however, Euler in fact discovered
the very opposite of what he was looking for – instead of finding what distinguished
one polyhedron from another, he found what they all had in common, namely the
relation (15.1). In modern terms we say that the Euler characteristic of all such
polyhedra is 2.

We elaborate on Pólya's dream in Section 15.2; we also refer there briefly to the
relation of the Euler characteristic to the Descartes *total angular defect*. Then in
Section 15.3 we show how Pólya's dream about Euler's aspirations has, in a sense,
come true. More precisely, by enlarging the concept of polyhedron to include all
closed surfaces, the Euler characteristic takes on the role assigned to it in the dream –
it does succeed, most admirably, in distinguishing between topologically distinct
surfaces.[1]

Of course, generalization doesn't occur in mathematics just to make dreams
come true. The concept of polyhedron certainly does not reach its fullest scope
with any 2-dimensional configuration; and so we give, in Section 15.4, an idea of
how the concept of polyhedron has grown in the hands of modern topologists, and
how the Euler characteristic has itself been adapted to fit into this broader concept.
Here one must mention the name of the great French mathematician Henri Poincaré
(1854–1912), a pioneer of modern topology, who understood the role of the **Betti
numbers** of a polyhedron, in connection with solutions of systems of differential
equations, and their topological invariance. The more refined version of the Euler
characteristic, suitable for any polyhedron of any dimension, is referred to as the
Euler–Poincaré characteristic, to mark Poincaré's contribution.

15.2 Pólya's dream

We begin by quoting from [69] where Pólya, in trying to show how Euler might
have discovered his famous formula (15.1), was "telling it as he would like to
believe it happened." At a certain point we deviate slightly from his actual account
in detail (though not in spirit) in order to present his story with what we believe to
be the fewest polyhedral examples possible to achieve the desired end.[2] Finally, we
describe the Descartes total angular defect Δ of a convex polyhedron. Descartes
used spherical trigonometry to prove that $\Delta = 4\pi$ for any such polyhedron. Without
giving a proof of this or of (15.1), we give a modified version of Pólya's proof

---

[1] One may remark (with tongue in cheek!) that we have here a beautiful example of the Marxist dialectic in
action. We want to distinguish polyhedra, but the Euler characteristic *negates* this by referring to what they all
have in common. So we enlarge the concept of polyhedron and thus *negate the negation*. Marxism seems to be
dead as a political philosophy – but might it be rising phoenix-like from the ashes?

[2] Pólya used 6 polyhedra and we use 5. Can any reader find a suitable sequence with fewer than 5 polyhedra?

that Euler's formula for polyhedra and Descartes' theorem are equivalent. This remarkable result appears in [70] but, in fact, it came to us in the form of lecture notes, taken by our late friend Dave Logothetti when he attended a talk Pólya gave at Stanford in March, 1974. Thus it is reasonable to argue that, in a sense, Descartes already knew the Euler formula.

Now let Pólya speak for himself. He wrote (see [69]):

In the "commentatio" (Note presented to the Russian Academy) in which his theorem on polyhedra (on the number of faces, edges and vertices) was first published, Euler gives no proof. In place of proof, he offers an inductive argument: He verifies the relation in a variety of special cases. There is little doubt that he also discovered the theorem, as many of his other results, inductively. Yet he does not give a direct indication of how he was led to his theorem, of how he "guessed" it, whereas in some other cases he offers suggestive hints about the ways and motives of his inductive considerations.

How was Euler led to his theorem on polyhedra? I think that it is not futile to speculate on this question although, of course, we cannot expect a conclusive answer.... One can imagine various approaches to the discovery (rediscovery) of Euler's theorem. I have presented two different approaches on former occasions.[3] I offer here a third one which, I like to think, could have been Euler's own approach.

1. Analogy suggests a problem. There is a certain analogy between plane geometry and solid geometry which may appear plausible even to a beginner. A circle in the plane is analogous to a sphere in space; the area enclosed by a curve is analogous to the volume enclosed by a surface in space; polygons enclosed by straight sides in the plane are analogous to polyhedra enclosed by plane faces in space.

Yet there is a difference. If we look closer the geometry of the plane appears as simpler and easier whereas that of space appears as more intricate and more difficult. We have a simple classification of polygons according to the number of their sides. ...

So, dreamt Pólya, Euler might have sought to classify polyhedra. First, he might have counted faces.

Here we deviate from Pólya by giving Figure 15.1, using the same polyhedra as Pólya in the first 4 cases, but our own fifth polyhedron is simply Pólya's third, the triangular prism, with one of its vertical faces dissected into 2 triangular faces. One should think of Figure 15.1 appearing initially without the entries of V and E along the top and without any of the values of V, E, and F which are shown.[4]

Now we try to see if the number of faces would classify the polyhedra. So beginning at the top of the table we enter F and fill in the values until we reach

---

[3] See [67], vol 1, pp. 35–43 and [70], vol 2, pp. 149–156, and also the annexed problems and solutions in both books. There is also a stimulating discussion of Euler's formula by Lakatos, and its proof, in [51].

[4] We prefer to show our table with the 0-dimensional objects (V), 1-dimensional objects (E), and 2-dimensional objects (F) appearing in that order, from left to right along the top. Although neither Euler nor Pólya did it this way, we argue that it prepares the ground for the generalization of the Euler characteristic to higher dimensions, and also that it allows us to formulate Euler's result in such a way that we can see precisely *what* is invariant.

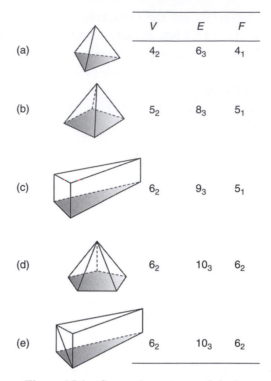

		V	E	F
(a)		4_2	6_3	4_1
(b)		5_2	8_3	5_1
(c)		6_2	9_3	5_1
(d)		6_2	10_3	6_2
(e)		6_2	10_3	6_2

Figure 15.1   Some elementary polyhedra.

an unsatisfactory result (these values have been given the subscript 1 because they were the first to be entered). In fact, we quickly see that although polyhedron (b) and polyhedron (c) have the same number of faces they are not equivalent.[5]

Continuing, from [69] again:

Here emerges a problem: Let us devise a classification of polyhedra that accomplishes something analogous to the simple classification of polygons according to the number of their sides. Yet in the case of polyhedra taking into account just the number of faces is not enough as the example of Figure 1 shows. . . .

What should we do to answer this question? Survey as many different forms of polyhedra as we can and count their faces and vertices?

We now go back to the top of the table in Figure 15.1 and enter V, then we begin to fill in the missing values for F and V (these values have the subscript 2) until we first reach an unsatisfactory result. What we see is that $(V, F) = (6, 6)$ for the polyhedra (d) and (e). But, again, these polyhedra are not equivalent.

---

[5] Pólya said they were *morphologically* different, adding "I am intentionally avoiding the standard term which, by the way, did not exist in Euler's time. One of the ugliest outgrowths of the 'new math' was the premature introduction of technical terms."

What's left besides faces and vertices? Of course – edges.[6] Again we go back to the top of the table in Figure 15.1 and enter E, then we fill in the missing values for E (these values have the subscript 3). Alas, this does not work either! As Pólya said:

They agree also in the number of edges, as they have agreed in the number of faces and vertices. And exploring further cases we find invariably: If two polyhedra have the same F and V, they also have the same E. Thus the number of edges contributes nothing to the classification of polyhedra over and above what the faces and vertices have done already. What a disappointment!

Yet there is something else. If the number E of edges is determined by the numbers F and V... then E is a function of F and V. Which function? Is it an increasing function? Does E increase wherever F increases? Does E necessarily increase with V?... Such or similar questions may lead to more examples ... and eventually to the guess

$$E = F + V - 2.$$

An unexpected, extremely simple relation, unique of its kind. What a triumph!

Pólya was here discussing his dream of how Euler might have discovered his wonderful formula and how it might have been *first* formulated, so that it is not surprising that it doesn't appear in exactly the same form as (15.1).

Now let us turn to a seemingly different aspect of polyhedra. Let us begin with a convex polyhedron P, homeomorphic to S^2 (meaning that P is sphere-like, in the sense that if the surface were made of rubber and inflated it would assume the shape of a sphere). Euclid proved that the sum of the face angles at any vertex of P is less than 2π; the difference between this sum and 2π is called the **angular defect** (or **angular deficiency**) at that vertex. If we sum the angular defects over all the vertices of P we obtain the **total angular defect** Δ; René Descartes proved that $\Delta = 4\pi$ for every convex polyhedron P. Thus, for example, there are 8 identical vertices on the cube and the angular defect at every vertex is $\frac{\pi}{2}$, so that the total angular defect Δ of the cube is 4π.

We now repeat Pólya's line of reasoning for those who might have missed it in Section 10.3, to show that $V - E + F = 2$ and $\Delta = 4\pi$ are equivalent statements. In the course of doing so, we obtain a result that takes us far beyond the domain of convex polyhedra. In fact, what we will prove is that

$$\Delta = 2\pi(V - E + F), \tag{15.2}$$

for any *closed rectilinear surface*. This immediately establishes the equivalence of the two formulas $V - E + F = 2$ and $\Delta = 4\pi$ in the case of convex polyhedra

---

[6] According to Pólya, "Euler was the first to introduce the concept of the 'edge of a polyhedron' and to give it a name (*acies*)... . Perhaps Euler introduced edges in the hope of a better classification, and we follow his example here."

homeomorphic to S^2. Let S be the total number of sides of the faces of our surface.[7] Then, since every edge of the surface is a side of exactly 2 faces,

$$S = 2E. \tag{15.3}$$

Now let P be a polyhedron homeomorphic to S^2 subdivided into V vertices, E edges, and F faces, so that every edge is incident with exactly 2 faces. Number the vertices $1, 2, \ldots, V$ and let the sum of the plane face angles at the nth vertex be σ_n. Then the angular defect at the nth vertex is

$$\delta_n = 2\pi - \sigma_n. \tag{15.4}$$

Note that δ_n will be positive if P is convex, but that, in general, δ_n may be negative or zero. Let

$$\Delta = \sum_{n=1}^{V} \delta_n. \tag{15.5}$$

We are now ready to prove (15.2). What we do is count the *sum of all the face angles* (which we call A) in two different ways. We first count by vertices. Then

$$A = \sum_{n=1}^{V} \sigma_n = \sum_{n=1}^{V} (2\pi - \delta_n) = 2\pi V - \Delta. \tag{15.6}$$

We next count by faces. Now if a face has m sides, then the sum of the interior angles of that face is $(m - 2)\pi$. Thus, if our polyhedron has F_m m-gons among its faces, those m-gons contribute $(m - 2)F_m\pi$ to the sum of the face angles. We thus arrive at the key formula

$$A = \sum_{m} (m - 2)F_m\pi = \left(\sum_{m} mF_m - 2\sum_{m} F_m \right)\pi. \tag{15.7}$$

Now

$$F = \sum_{m} F_m. \tag{15.8}$$

Also, since each m-gon has m sides, the contribution to the number of sides from the m-gonal faces is mF_m. Thus

$$S = \sum_{m} mF_m. \tag{15.9}$$

---

[7] It is very important to the understanding of this proof to distinguish carefully between the meaning of a *side* and that of an *edge*. Unfortunately, this is often not done. It is even common to confuse "side" and "face."

We put together (15.7), (15.8) and (15.9) to infer that

$$A = (S - 2F)\pi. \tag{15.10}$$

Comparing (15.10) with (15.6) we conclude that $2\pi V - \Delta = (S - 2F)\pi$, or that

$$\Delta = \pi(2V - S + 2F). \tag{15.11}$$

But, of course, if $S = 2E$, we obtain (15.2).

Remarks

(i) Pólya obtained (15.2) without introducing either S or A. However, by introducing these terms, we arrive at the more general equation (15.11). We thereby see that Pólya's argument immediately generalizes, not only to arbitrary 2-dimensional closed surfaces, which need not even be orientable (for details see [25]), but even to arbitrary 2-dimensional polyhedra in the most general possible sense.

(ii) Grünbaum and Shephard proposed the excellent idea of a dual for Descartes' theorem for polyhedra in [22]). The introduction of S along with its dual R (for the number of rays) and (15.11) resulted in a perfectly general formulation of that duality (see [34] for details). In fact, while $\Delta = 2\pi\chi$ is only true for closed surfaces, $\Delta' = 2\pi\chi$ (where $\Delta' = \pi(2F - R + 2V)$) is true for *any* 2-dimensional polyhedron in the most general sense.

(iii) René Descartes (1596–1650) and Leonhard Euler (1707–1783) worked on these subjects independently – yet, as we have seen, George Pólya (1887–1985) has shown, in an elementary fashion, that their seemingly different, but equally profound, formulas for convex polyhedra homeomorphic to S^2 (polyhedra that can be deformed to coincide with the surface of a sphere) are entirely equivalent to each other. It is a trivial theorem that Descartes did not know about Euler's work. It is a less obvious theorem that Euler could not have known about Descartes' work – since Descartes' work on this matter [13] was not printed until a century after Euler's death.

(iv) The problem that Pólya believed Euler set out to solve remains unsolved to this day! Nevertheless . . .

*15.3 ... The dream comes true

So Euler could not classify polyhedra by the simple method Pólya envisaged in his dream; instead he found the property, which we have already mentioned,

$$V - E + F = 2, \tag{15.12}$$

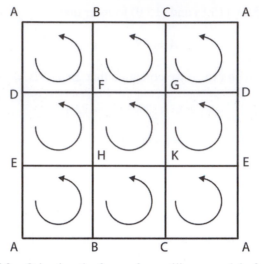

Figure 15.2   Orienting the faces of a rectilinear model of the torus.

common to *all* polyhedra homeomorphic to S^2. Is that the end of the story? Certainly not! For topologists have found it useful – indeed, necessary – to enlarge the concept of polyhedron, and then the quantity $V - E + F$, which we call the **Euler characteristic**, does serve to distinguish between types of polyhedra.

The first generalization we will adopt is, as you would expect, that of a **closed orientable surface**. Thus we now consider topological spaces[8] S such that each point p of S has a neighborhood homeomorphic to an open disk. The condition of orientability is equivalent to the requirement that S should be embeddable in $\mathbb{R}^3$. We may always realize S by a rectilinear model, so that S consists of vertices, edges, and faces. Then the condition of orientability asserts that each face may be oriented so that a common side of 2 faces receives *opposite* orientations from those 2 faces. A typical, and important, example of a closed orientable surface is the torus; a rectilinear model of the torus, with its faces coherently oriented, is given in Figure 15.2.

By the use of homology theory, one may show that the Euler characteristic χ is a topological invariant; that is, if two surfaces S_1 and S_2 are homeomorphic, then $\chi(S_1) = \chi(S_2)$. This is really very remarkable, since $\chi(S)$ is defined combinatorially, using a subdivision of S into faces, edges, and vertices. Nevertheless $\chi(S)$ depends only on the underlying topology of S and not on the combinatorial structure imposed on S.

This result explains why (15.12) holds for all polyhedra homeomorphic to S^2, for such spaces would all have had the same Euler characteristic. That $\chi = 2$ for

---

[8] Of course, S no longer designates the number of sides!

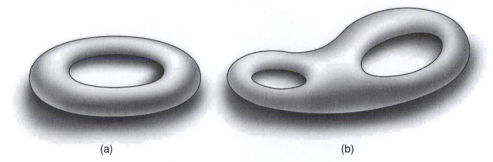

Figure 15.3   (a) $g = 1$ (the torus); (b) $g = 2$ (the double torus).

such spaces follows from the proof of the topological invariance of χ. The basic
theorem is that

$$\chi = p_0 - p_1 + p_2, \qquad (15.13)$$

where p_i is the **ith Betti number** of the space. Let us give an indication of what
that means. With any topological space K – but let us think of a (finite) rectilinear complex for simplicity – we may associate certain abelian groups $H_r K$,
$r = 0, 1, 2, \ldots,$ called the **homology groups** of K, which, roughly speaking,
count the r-dimensional "holes" in K. If the space is n-dimensional we only
have homology groups up to dimension n. Then p_r is the **rank** of $H_r K$. Now for
a closed orientable surface S_g of **genus** g, that is, with g holes or g handles (see
Figure 15.3), the Betti numbers are given by $p_0 = 1$, $p_1 = 2g$, $p_2 = 1$, so that

$$\chi(S_g) = 2 - 2g. \qquad (15.14)$$

Of course, the sphere S^2 has no holes, $g = 0$, so $\chi(S^2) = 2$, explaining the results
of Section 15.2 embodied in (15.12). We see from (15.14) that $\chi(S)$ can take as
values any even number no greater than 2. Thus Euler's aspirations are realized,
the dream comes true; $\chi(S)$ *does* distinguish between various closed orientable
surfaces. It takes different values on non-homeomorphic surfaces, and is, in fact,
entirely determined by the genus of S, according to formula (15.14).

Can we generalize further? Well, we can drop the requirement of orientability.
We then obtain a family of closed surfaces characterized by the number k of cross-
caps (Möbius bands) inserted into the sphere. The case $k = 1$ is the best known;
we then have the real projective plane $\mathbb{R}P^2$ (see Figure 15.4).

Now formula (15.13) holds for any compact 2-dimensional polyhedron, in the
broadest possible sense. For a non-orientable surface $S^{(k)}$ with k cross-caps, we
have $p_0 = 1$, $p_1 = k - 1$, $p_2 = 0$, so that

$$\chi\left(S^{(k)}\right) = 2 - k; \qquad (15.15)$$

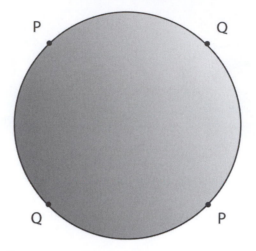

Figure 15.4 $\mathbb{R}P^2$, represented as a circular disk with diametrically opposite points on the boundary identified.

in particular

$$\chi\left(\mathbb{R}P^2\right) = 1. \tag{15.16}$$

From (15.15) we see that we can realize any integer no greater than 2, even or odd, as the Euler characteristic of a closed surface, by allowing non-orientable surfaces. However, we now lose the capacity of χ to specify topological type, since an orientable surface of genus g and a non-orientable surface with $2g$ cross-caps have the same Euler characteristic.

Can we achieve any integer as the Euler characteristic of a 2-dimensional polyhedron? The answer is certainly yes, provided that we further broaden the concept of polyhedron. We have already slipped in a reference to a (finite) rectilinear complex K, that is, a space broken up into vertices, edges, and faces, where a (closed) face is simply a polygonal region. Then it is plain that, for any positive integer $n \geq 2$, we can construct a 2-dimensional polyhedron K with $\chi(K) = n$ by taking K to be a bunch of $(n-1)$ balloons (see Figure 15.5). For then $p_0 = 1$, $p_1 = 0$, $p_2 = n - 1$.

*15.4 Further generalizations

An obvious generalization of a closed orientable surface is obtained by dropping the requirement that our space be 2-dimensional. We are thus led to the concept of a ***closed orientable manifold*** of dimension n. Just as S^2 was the first case of a closed orientable surface to be studied, so is S^n the natural starter for a study of closed orientable n-manifolds. Indeed, Schläfli [73] generalized (15.12) to show,

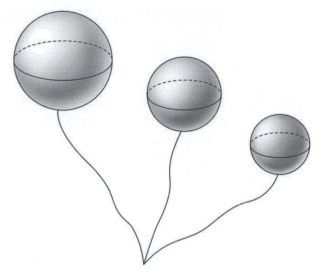

Figure 15.5   A bunch of n balloons ($n = 3$).

in effect, that

$$\chi(S^n) = \begin{cases} 2, & n \text{ even} \\ 0, & n \text{ odd.} \end{cases} \qquad (15.17)$$

Here we regard χ as the alternating sum

$$\chi = \sum_{r=0}^{n} (-1)^r \alpha_r, \qquad (15.18)$$

where α_r is the number of r-dimensional cells in some cellular decomposition of the manifold. Thus (15.17) generalizes (15.12); the corresponding generalization of (15.13), which proves the topological invariance of χ, is

$$\chi = \sum_{r=0}^{n} (-1)^r p_r, \qquad (15.19)$$

where p_r is the rth Betti number, as before. It is customary to call χ, so generalized, the **Euler–Poincaré characteristic**, since Poincaré proved the topological invariance of the Betti numbers. (It appears that we owe to Emmy Noether the very important observation that we should be dealing with certain *abelian groups*, namely the homology groups, rather than with *numbers*, which are merely their ranks. The homology groups are the true topological invariants.) The proof of the equivalence of (15.18) and (15.19) is a nice exercise in linear algebra.

Of course, (15.17) follows immediately from (15.19) since, for the sphere S^n,

$$p_0 = p_n = 1;$$
$$p_r = 0, \quad r \neq 0, n; \tag{15.20}$$

however, Schläfli used basically combinatorial arguments in [73].

We observe that the equivalence of (15.18) and (15.19) is valid for *any n*-dimensional polyhedron in the most general sense. We may, if we like, confine attention to those spaces admitting the structure of an n-dimensional simplicial complex K. Such a complex K is a collection of **simplexes** of dimension $0, 1, \ldots, n$, where an r-simplex is the convex hull of a set of $(r + 1)$ independent points. Thus

> a   0-simplex is a vertex,
>
> a   1-simplex is an edge,
>
> a   2-simplex is a triangle,
>
> a   3-simplex is a tetrahedron, ...

Moreover, distinct simplexes intersect (if at all) in a common face of both. Since the Euler–Poincaré characteristic is a topological invariant, we may, of course, define it for any space homeomorphic to the underlying space of a finite simplicial complex – thus, in particular, for a genuine, geometric n-sphere. If we define a (compact) polyhedron as such a homeomorph, we have achieved a very substantial generalization, taking us as far as we would wish to go in this direction in this chapter.

Let us point to just one obvious advantage of having generalized the Euler characteristic as we have done. Given two spaces X and Y we can form their topological product $X \times Y$. Now if X, Y are compact polyhedra one may show that $X \times Y$ is also a compact polyhedron. It is then natural to ask for the connection between $\chi(X)$, $\chi(Y)$ and $\chi(X \times Y)$. The answer is delightfully simple (to those familiar with the foundations of homology theory)!

Theorem 15.1 $\chi(X \times Y) = \chi(X)\chi(Y)$.

The proof is no less pleasing (in our judgment). As one seeks to calculate the homology groups of $X \times Y$ in terms of those of X and Y, one finds a very simple formula if one confines oneself to the case of homology groups with rational coefficients $\mathbb{Q}$ (or, indeed, with any *field* of coefficients replacing $\mathbb{Q}$), namely,

$$H_r(X \times Y; \mathbb{Q}) = \bigoplus_{s+t=r} H_s(X; \mathbb{Q}) \otimes H_t(Y; \mathbb{Q}) \tag{15.21}$$

(recall that these homology groups are, in fact, vector spaces over $\mathbb{Q}$). From (15.21) we immediately infer that

$$p_r(X \times Y) = \sum_{s+t=r} p_s(X)p_t(Y).$$ (15.22)

There is a very nice way to express (15.22). Let us consider the formal polynomial $\sum_{r \geq 0} p_r(X)x^r$. We call this the **Poincaré polynomial** of X, and write it $P_X(x)$. Then (15.22) simply asserts that

$$P_{X \times Y}(x) = P_X(x)P_Y(x).$$ (15.23)

This almost completes the proof of Theorem 15.1 – and really explains why it is true. For plainly

$$\chi(X) = P_X(-1).$$ (15.24)

So Theorem 15.1 follows immediately from (15.23).

Theorem 15.1 could not have been brought into existence without the generalization of the Euler characteristic; moreover, it gives us great confidence that we've chosen the *right* generalization. Finally, our proof, like any good proof, leads us smoothly into further questions which, unfortunately, we cannot allow ourselves to be explicit about here.

We have not attempted in this chapter to bring the story right up to date – research continues today on modified versions and adaptations of the Euler–Poincaré characteristic. The progress seems, in some sense, to be inevitable – there is always, somewhere, a George Pólya having a dream![9]

---

[9] The content of this chapter originally appeared in *The American Mathematical Monthly* [40].

16

Tying some loose ends together – Symmetry, group theory, homologues, and the Pólya enumeration theorem

16.1 Symmetry: A really big idea

The concept of symmetry plays a strong role today in many of the exact sciences. Thus, for example, theoretical physicists, in searching for a unified field theory, have been led to the notion of *supersymmetry*, applied to the (super)strings, which, as some believe, are the fundamental building blocks of the Universe. Perhaps the foremost exponent of this position is the American physicist Edward Witten, of the Princeton Institute of Advanced Study, who, a few years ago, won a Fields Medal – the most prestigious award that can be given to a mathematician[1] – for his fundamental theoretical contributions to superstring theory. Even more recently (August, 1998, at the International Congress of Mathematicians held in Berlin) the Cambridge mathematician Richard Borcherds was awarded a Fields Medal for his contribution to the development of symmetry theory, especially with respect to Witten theory and its relation to the sporadic finite groups.

What, then, *is* symmetry? In this chapter we attempt to make the idea precise, keeping our applications of the concept at a level where, as we hope, they will be appreciated by our readers. We will confine ourselves to the use of the symmetry concept within mathematics; and we must first of all emphasize that notions of symmetry, while fundamental to geometry, are certainly, and importantly, to be found in areas of mathematics outside geometry. Thus, for example, any function of two variables, $f(x, y)$, may be described as **symmetric** if $f(x, y) = f(y, x)$; and similarly, any function of three variables, $f(x, y, z)$, may be described as **symmetric** if its value remains unchanged under any permutation (there are 6 if one includes the identity permutation) of the variables x, y, z. As we have written about

---

[1] It is a curious fact that only two mathematicians have received Nobel Prizes for *mathematical* work. The Nobel Prize in Economics has been awarded twice to Mathematicians: John Nash in 1994 and Robert Aumann in 2005. There is no Nobel Prize for mathematics, but there is now an Abel Prize for mathematics, awarded annually by the Norwegian Academy of Sciences.

elsewhere (see [24]) the binomial coefficient $\binom{n}{r}$ becomes a symmetric function if one regards it as a function of r and s, where $r + s = n$. Thus the relation $\binom{n}{r} = \binom{n}{n-r}$ takes the *symmetric* form

$$\binom{n}{r \ s} = \binom{n}{s \ r}$$

if one writes $\binom{n}{r \ s}$, with $r + s = n$, instead of the (apparently) simpler $\binom{n}{r}$. Similarly the trinomial coefficient $\binom{n}{r \ s \ t}$, $r + s + t = n$, is a symmetric function of the variables[2] r, s, t.

As another example, symmetric functions of the roots of polynomial equations

$$x^n + a_1 x^{n-1} + \cdots a_{n-1}x + a_n = 0$$

play an essential role in the algebra of polynomials. We know that a polynomial equation of degree n over the complex numbers has n roots. A crucial theorem states that if $\alpha_1, \alpha_2, \ldots, \alpha_n$ are the roots, then any symmetric polynomial in $\alpha_1, \alpha_2, \ldots, \alpha_n$, with integer coefficients may be expressed, uniquely, as a polynomial in $a_1, a_2, \ldots, a_n$ with integer coefficients. Thus, for example, with $n = 3$, we have

$$\alpha_1 + \alpha_2 + \alpha_3 = -a_1,$$
$$\alpha_1^2 + \alpha_2^2 + \alpha_3^2 = a_1^2 - 2a_2,$$
$$\alpha_1^3 + \alpha_2^3 + \alpha_3^3 = -a_1^3 + 3a_1a_2 - 3a_3.$$

You may wish to check the above formulas, using the identity

$$x^3 + a_1x^2 + a_2x + a_3 = (x - \alpha_1)(x - \alpha_2)(x - \alpha_3).$$

(If this seems too difficult check them first in a simple particular case like $\alpha_1 = 1, \alpha_2 = 2, \alpha_3 = 3$.) The sequence above may be continued by expressing $\alpha_1^4 + \alpha_2^4 + \alpha_3^4$ in terms of a_1, a_2, a_3, obtaining

$$\begin{aligned}
\alpha_1^4 + \alpha_2^4 + \alpha_3^4 &= (\alpha_1^2 + \alpha_2^2 + \alpha_3^2)^2 - 2(\alpha_1^2\alpha_2^2 + \alpha_1^2\alpha_3^2 + \alpha_2^2\alpha_3^2) \\
&= (a_1^2 - 2a_2^2)^2 - 2\{(\alpha_1\alpha_2 + \alpha_1\alpha_3 - \alpha_2\alpha_3)^2 - 2\alpha_1\alpha_2\alpha_3(\alpha_1 + \alpha_2 + \alpha_3)\} \\
&= a_1^4 - 4a_1^2a_2 + 4a_2^2 - 2(a_2^2 - 2a_1a_3) \\
&= a_1^4 - 4a_1^2a_2 + 4a_1a_3 + 2a_2^2.
\end{aligned}$$

It's an interesting fact that the above theorem (about the roots of a symmetric polynomial) maybe used to prove that if F_n is the nth Fibonacci number, then

---

[2] In order to stress the symmetry, the eminent French mathematician Henri Cartan (1904–2008) employed the simple, but revealing, notations $(r \ s)$, $(r \ s \ t)$ for binomial and trinomial coefficients, respectively.

$F_m \mid F_n$ if $m \mid n$; and that, if L_n is the nth Lucas number then $L_m \mid L_n$ if $m \mid n$ with odd quotient (we may then say that $m \mid n$ *oddly*).

However, our main purpose in this chapter will, in fact, be to explain the nature and role of symmetry in *geometry*. It is our belief that its role in other parts of mathematics will then become easier to understand. Interestingly, it turns out that to describe the nature of symmetry in geometry, it is necessary to introduce some basic concepts in a very important part of algebra known as **group theory**. Thus, in Section 16.2, we describe these concepts and explain how they enable us to define the key notion of the *symmetry group* of a geometric configuration A. Let us add that it is not at all surprising that we need basic concepts from group theory in geometry: for the great German mathematician Felix Klein (1849–1925), in his famous Erlanger Programm, *defined* geometry in terms of the group of allowed transformations of a given set of points; and we will be adopting Klein's point of view. For example, the **Euclidean geometry** of the plane $\mathbb{R}^2$ is to be understood as the set of properties of configurations in $\mathbb{R}^2$ invariant under the group of transformations of $\mathbb{R}^2$ generated by *translations, rotations, and reflections*.

In Section 16.3, we use the group theory presented in Section 16.2 to give a definition of the concept of *homologue*, due to the great mathematician and expositor George Pólya (1887–1985). Pólya never wrote this idea down, but asked the two of us (PH and JP), near the end of his life, to do so on his behalf. In fact, our interpretation of Pólya's idea makes that idea applicable in a situation more general than that which he discussed with us, that is, whenever one considers groups acting on sets – though the most vivid examples are going to come from geometry.

In Section 16.4 we describe one of the greatest theorems of combinatorics, the celebrated **Pólya enumeration theorem**[3]. This theorem is very often applied in a geometric context (our own applications will be geometric); however, once again, it is not in essence a theorem of geometry at all, but rests on the foundations of the idea of a (finite) group acting on a finite set – a basic idea of combinatorics.

Section 16.5 is a treatment of the idea of *odd* and *even* permutations. Those who would have preferred to get their group theory in one piece rather than two may, of course, read this section immediately after they have studied Section 16.2; but we thought that, though the basic idea of the section plays an important role in the study of polyhedral symmetry, it might be rather indigestible if offered to the reader before much of the geometry had been treated.

Section 16.6 is very different in nature from the rest of the chapter. As we have said, the two of us knew George Pólya well, for many years, and we would like to

---

[3] This theorem is described in *Handbook of Applicable Mathematics*, edited by Ledermann [52], as the Redfield–Pólya theorem.

take this opportunity to say a few personal words about this extraordinary man. We believe our readers will enjoy getting to know him, even at second hand.

*16.2 Symmetry in geometry

Although, as we have said, the concept of symmetry may be found in all parts of mathematics, and in all those areas of science to which mathematics makes an essential contribution, it remains true that its best known applications are in geometry. One finds the idea of symmetry frequently referred to in the best of the most elementary treatments of geometry [83], and also in the best of the treatments of geometry suitable for those making a serious study of mathematics at the university level, for example [77]. But many elementary treatments of the symmetry of geometric figures are confusing and misleading, largely because those treatments never make it clear what geometry is (nor what symmetry is!).

So we want first to give a precise definition of **geometry**; of course, we are not advocating that this be done the way we are doing it in teaching elementary students, but it does seem to us appropriate to introduce these fundamental mathematical ideas to the readers of our book. We maintain, with Klein, that no treatment of *geometry* is complete without the idea of *symmetry*; and that no clear idea of symmetry is possible without the basic notion of a *group*.

Thus we start this discussion with a definition.

Definition of a group

Let G be a set and let $*$ be a binary operation on G; that is, given two elements g, h of G, then $g * h$ is itself an element of G. We then say that $(G, *)$ is a **group** (often abbreviated to "G is a group" if $*$ may be understood) if it satisfies the following 3 axioms:

I (associative law) for all g, h, k in G, $(g * h) * k = g * (h * k)$;

II (existence of 2-sided identity) there exists an element e in G such that

$$g * e = e * g = g, \quad \text{for all } g \text{ in } G;$$

III (existence of 2-sided inverses) there exists, to each g in G, an element $\bar{g}$ in G such that

$$g * \bar{g} = \bar{g} * g = e.$$

Let us immediately give two examples.

Example 16.1 Let G be the set of integers (positive, negative, and zero) and let $g * h$ mean $g + h$, the ordinary addition of integers. Then G is a group, with $e = 0$ and $\bar{g} = -g$.

Example 16.2 Consider the set S of permutations of the set $\{1, 2, 3\}$. There are 6 such permutations; we specify each permutation s by saying where s sends 1, 2, 3. Thus

s_1: $1 \to 1, 2 \to 2, 3 \to 3$; s_2: $1 \to 1, 2 \to 3, 3 \to 2$; s_3: $1 \to 2, 2 \to 1, 3 \to 3$;
s_4: $1 \to 2, 2 \to 3, 3 \to 1$; s_5: $1 \to 3, 2 \to 1, 3 \to 2$; s_6: $1 \to 3, 2 \to 2, 3 \to 1$.

The binary operation $s_i * s_j (1 \le i, j \le 6)$ simply produces the permutation s_i followed by the permutation s_j. It is obvious that the associative law is satisfied.[4] The identity permutation is s_1. Finally, the inverses are as follows:

The inverse of s_1 is s_1

The inverse of s_2 is s_2

The inverse of s_3 is s_3

The inverse of s_4 is s_5

The inverse of s_5 is s_4

The inverse of s_6 is s_6.

It is customary to write, e.g. $\left(\begin{smallmatrix} 1 & 2 & 3 \\ 2 & 1 & 3 \end{smallmatrix}\right)$, $\left(\begin{smallmatrix} 1 & 2 & 3 \\ 3 & 1 & 2 \end{smallmatrix}\right)$ for the permutations s_3, s_5.

Notice what these two examples have in common, and where they differ. Example 16.1 is an *infinite* group, and the group operation is *commutative*, that is

$$g * h = h * g, \quad \text{for all } g, h \text{ in } G.$$

Example 16.2 is a *finite* group, and the group operation is not commutative, thus

$$s_2 * s_3 = s_4, \qquad s_3 * s_2 = s_5.$$

In both cases, the group has only one identity element, and each element has only one inverse; moreover, if $\bar{g}$ is the inverse of g, then g is the inverse of $\bar{g}$.

As you may verify in Examples 16.1 and 16.2:

(a) there is only one identity element;
(b) each element has only one inverse;
(c) if $\bar{g}$ is the inverse of g, then g is the inverse of $\bar{g}$.

In fact, (a), (b), and (c) are true in *any* group G.

Where the group is finite, we can display the entire group law by a square ***group table***. Thus, for Example 16.2, the group table is

---

[4] When the binary operation $g * h$ says, "Do g, then do h," it is always associative. (Do you understand why?)

$*$	s_1	s_2	s_3	s_4	s_5	s_6
s_1	s_1	s_2	s_3	s_4	s_5	s_6
s_2	s_2	s_1	s_4	s_3	s_6	s_5
s_3	s_3	s_5	s_1	s_6	s_2	s_4
s_4	s_4	s_6	s_2	s_5	s_1	s_3
s_5	s_5	s_3	s_6	s_1	s_4	s_2
s_6	s_6	s_4	s_5	s_2	s_3	s_1

(We read off $s_i * s_j$ by looking in the row i and column j.) Notice that each row and each column is a permutation of the list of group elements. This is so for any finite group (sometimes this information can be used to complete a table without working out all the individual cases). The **order** of a group is the number of elements in the group, so the order is infinite (Example 16.1) or some finite number (in the case of Example 16.2 the order is 6).

Just so that you will feel reassured that we have not forgotten our purpose in introducing you to the idea of a group, let us tell you that the group of Example 16.2 is called the **symmetric group** on 3 symbols, and written S_3. It will certainly reappear!

But before we turn to geometry, we require one more result from group theory. This result, due to the great French mathematician Joseph-Louis Lagrange (1736–1813), relates the order of a finite group G and the order of a *subgroup H of G*.

Now if $(G, *)$ is a group and H is a subset of G, we say that H is a **subgroup** of G if $g_1 * g_2$ belongs to H whenever g_1, g_2 belong to H, and if the induced binary operation on H is a group operation. For this last requirement to be satisfied, it is necessary and sufficient for the identity element to belong to H, and for the inverse of any element in H to belong to H.

Example 16.1 (revisited) We have the additive group $\mathbb{Z}$ of integers. The subset consisting of even integers is a subgroup, usually written $2\mathbb{Z}$. The subset consisting of non-negative integers is closed under addition but is *not* a subgroup because the additive inverse of a positive integer is negative.

Before stating Lagrange's theorem, we introduce some important standard notation. It is customary to write a group G *multiplicatively*, especially if it is not commutative. That is, we write the group operation as multiplication and even (as in ordinary algebra) suppress the product symbol, so that $g_1 * g_2$ is simply written as $g_1 g_2$. Sometimes we may even go further and write the identity element as 1 instead of e; but we will not adopt this notation unless we judge there is no risk at all of confusion. We will, however, always write g^{-1} for the inverse of g. Notice that, since the associative law holds, the positive powers of g, that is $g, g^2, g^3, \dots,$

have an unambiguous meaning and so, indeed, do the negative powers if we define

$$g^{-n} = (g^n)^{-1}, \quad \text{for } n \text{ positive.} \tag{16.1}$$

Of course, (16.1) then also holds for n negative (just as in ordinary algebra).

Notice, too, that the inverse satisfies the basic law

$$(gh)^{-1} = h^{-1}g^{-1}, \quad g, h \text{ in } G. \tag{16.2}$$

Here we have been careful to state the law so that we do *not* require commutativity.

Now let G be a finite group of order n and let H be a subgroup of G. Of course H is also a finite group; let its order be m. Then we have the following important result.

Lagrange's Theorem *The order of H divides the order of G that is, m | n.*

Proof.[5] Let g be an element of G (we write: $g \in G$), and let gH be the set of all elements gh, as h ranges over the elements of H. We call the set gH a (left) **coset** of H in G and g a **representative** of the coset gH.

We now prove that, if $g_1 H$, $g_2 H$ are two (left) cosets, then either they are disjoint or they coincide. For if $g \in g_1 H$, then $g = g_1 h_0$ for some $h_0 \in H$, so, for all $h \in H$,

$$gh = g_1 h_0 h, \quad \text{with} \quad h_0 h \in H,$$

so $gh \in g_1 H$, and thus $gH \subseteq g_1 H$. But $g_1 = gh_0^{-1}$, with $h_0^{-1} \in H$, so we may repeat the argument with the roles of g and g_1 exchanged (and h_0^{-1} replacing h_0), concluding that $g_1 H \subseteq gH$, so that, finally,[6] $g_1 H = gH$. Thus if $g \in g_1 H \cap g_2 H$, then $g_1 H = gH = g_2 H$, establishing that $g_1 H$ and $g_2 H$ coincide if they are not disjoint.

We have next to prove that every coset gH has exactly the same number of elements as H itself. For consider the function φ from H to gH which sends h to gh. This function certainly maps H onto gH; for the elements of gH are exactly the elements gh, as h ranges over H. Moreover, φ is one-one; for if $gh_1 = gh_2$, then $h_1 = g^{-1}(gh_1) = g^{-1}(gh_2) = h_2$. Thus φ sets up a one-one correspondence between H and[7] gH.

Now suppose that H has m elements. Then every (left) coset has m elements; and G, as a set, is the disjoint union of a certain number of cosets. If, then, G is the disjoint union of k cosets, we have proved that

$$n = km, \tag{16.3}$$

establishing Lagrange's theorem.   □

---

[5] We advise you to read this argument carefully. It is really subtle.
[6] We have given you here an argument which does *not* depend on G or H being finite.
[7] See previous footnote.

Remarks

(i) The number k which appears in (16.3) is called the *index* of H in G.

(ii) Notice that we could have argued with right cosets Hg, so the index is also the number of disjoint right cosets. Obviously we would have arrived at the same relationship (16.3), so the index is also the number of disjoint right cosets. This is significant because it is by no means true in general that every left coset is a right coset.

(iii) One may set up a one-one correspondence between the left cosets of H in G and the right cosets of H in G – but one must be careful – there is no well-defined function sending the coset gH to the coset Hg.

This is as far as we need to take the group theory in order to be precise as to what we mean by a *geometry* on a configuration A, and the symmetry of the resulting geometric figure. However, some further (and obviously relevant) group theory will be found in Section 16.5.

We have been guided in our definitions by the approach of the great German mathematician Felix Klein (1849–1925) to explaining the nature of geometry. Consider, for example, the usual plane Euclidean geometry, in which we study the properties of planar figures which are invariant under certain *Euclidean motions*. These motions certainly include *translation* and *rotation* in the plane of the figure, but it is a matter of choice whether they include *reflection*. For example the FAT 7-gon of Figure 16.1(a) is invariant under rotations through $\frac{2\pi}{7}$ about its center, but not under reflection about any axis through its center. Thus, to define our geometry, we must decide whether we allow Euclidean motions which reverse orientation. Of course, if we allow certain Euclidean motions, we must also allow compositions and inverses of such motions, so we postulate a certain *group G* of allowed motions of the *ambient* plane, the plane containing our planar figure. If A is such a planar figure, then, for any $g \in G$, Ag is again[8] a planar figure and, *in the G-geometry of A*, we study the properties of the figure A which it shares with all the figures Ag as g varies over G; such properties are called the *G-invariants* of A, abbreviated to *invariants* if, but only if, the *group G* may be understood. It is the *group G* which, according to Klein, determines the *geometrical* nature of A.

Example 16.3 Let G be the group of motions of the plane generated by translations, rotations, and reflections (in a line); we call this the Euclidean group in 2 dimensions and may write it E_2. Then the Euclidean geometry of the plane is the study of the properties of subsets of the plane which are invariant under the motions in E_2. For example, the property of being a polygon is a Euclidean property; the number

---

[8] We prefer to write Ag for the image of A under the motion g, rather than gA. This is because, in the group G, gh means "first g, then h."

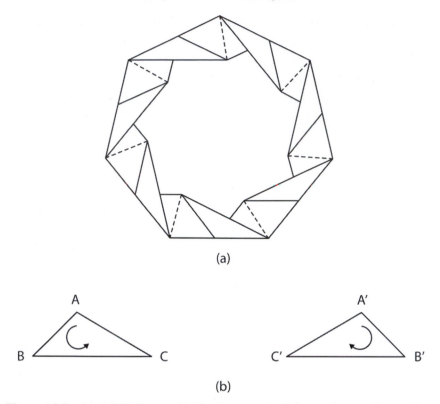

(a)

(b)

Figure 16.1   (a) A FAT 7-gon. (b) The figure on the left may be transformed by a rotation in 3 dimensions to the figure on the right, reversing the orientation of the triangle.

of vertices and sides of a polygon is a Euclidean invariant. On the other hand, as we have hinted, orientation is not invariant with respect to this group, though it would be if we disallowed reflections. Thus, by means of a motion in E_2 the triangle ABC may be turned over (flipped) to form the triangle A'B'C' as shown in Figure 16.1(b). But the orientation $\overrightarrow{ABC}$ is counterclockwise, while the orientation $\overrightarrow{A'B'C'}$ is clockwise.

We may step up a dimension, passing to the group E_3 of Euclidean motions in 3-dimensional space. Notice that it is natural to think of reflections in a line (of a planar figure) as a motion since it can be achieved by a rotation in some suitable ambient 3-dimensional space containing the plane figure. However, it requires a greater intellectual effort to think of reflection in a plane (of a spatial figure) as a motion in some ambient 4-dimensional space! Who would think of turning the golden dodecahedron of Section 9.4 inside out? Thus it is common not to include such reflections in defining 3-dimensional geometry. This preference is, of course, a consequence of our experience of living in a 3-dimensional world and has no mathematical basis. However, whenever we are highlighting the construction of actual physical models of geometrical configurations, it is entirely reasonable to omit motions to which the models themselves cannot be subjected.

We now move to a precise definition of symmetry. Let a geometry be defined on the ambient space of a configuration A by means of the group of motions G. Then the **symmetry group** of A, relative to the geometry defined by G, is the subgroup G_A of G consisting of those motions $g \in G$ such that $Ag = A$, that is, those motions which map A onto itself, or, as we say, under which A is **invariant**. Thus, for example, if our geometry is defined by rotations and translations in the plane, and if A is an equilateral triangle, then its symmetry group G_A consists of rotations about its center through $0°$, $120°$, and $240°$; if, in our geometry, we also allow reflections, then the symmetry group has 6 elements instead of 3, and is, in fact, the very well-known group S_3 (recall Example 16.2), called the **symmetric group on 3 symbols** – the symbols may be thought of in this case as the vertices of the triangle. We must repeat for emphasis that the symmetry group G_A of the configuration A is a *relative* notion, depending on the choice of geometry G.

It is plain that no compact (bounded) configuration can possibly be invariant under a non-zero translation. Thus when we are considering the symmetry group of such a figure we may suppose G to be generated by rotations and, perhaps, reflections. Moreover, any such motion in the plane is determined by its effect on 3 independent points and any such motion in 3-dimensional space is determined by its effect on 4 independent points. Since a (plane) polygon has at least 3 vertices and a polyhedron has at least 4 vertices, and since any element of the symmetry group of a polygon or a polyhedron must map vertices to vertices, it follows that the symmetry group of a polygon or a polyhedron is *finite* (compare the symmetry groups of a circle or a sphere).

The symmetry group of any polygon with n sides is, by the argument above, a subgroup of S_n, the group of permutations of n symbols, also called the **symmetric group** on n symbols. If G is generated by rotations alone, and the polygon is regular, its symmetry group is the cyclic group[9] of order n, often written C_n, generated by a rotation through an angle of $\frac{2\pi}{n}$ radians about the center of the polygonal region. If G also includes reflections, then the symmetry group has $2n$ elements and includes n reflections; this group is called the **dihedral** group of order $2n$ and is usually written D_n. Remember that D_n has $2n$ elements, not just n elements.

In discussing the symmetry groups of polyhedra, however, we will, as indicated earlier, always assume that the geometry is given by the group G generated by

---

[9] A cyclic group of order n is a group of order n, all of whose elements are powers of a given element g, called a **generator** of the group.

rotations in 3-dimensional space. Then the symmetry group of the regular tetrahedron is the so-called **alternating** group A_4. In general, A_n is the subgroup of S_n consisting of the even permutations[10] of n symbols; it is of index 2 in S_n, whose order is $n!$, so that its order is $\frac{1}{2}(n!)$. Thus the order of A_4 is 12. As we have already discussed in Section 10.2, the cube and the regular octahedron have the same symmetry group, namely S_4. It is easy to see why the symmetry groups are the same; for the centers of the faces of a cube are the vertices of a regular inscribed octahedron, and the centers of the faces of a regular octahedron are the vertices of an inscribed cube.[11] Likewise, and for the same reason, the regular dodecahedron and the regular icosahedron have the same symmetry group, which is A_5. It is a matter of interest and relevance here that the elements of the symmetry group of the diagonal cube (of Section 9.3) permute the four braided strips from which the model is made and thus the symmetry group is S_4.

We are now in a position to give at least one possible precise meaning to the statement "Figure A is more symmetric than Figure B." We assume that A and B are defined as geometric figures by the same group of motions G. If it happens that the symmetry group G_A of A strictly contains the symmetry group G_B of B, then we are surely entitled to say that A is more symmetric than B. Notice that the situation described may, in fact, occur because B is obtained from A by adding features which destroy some of the symmetry of A. For example each of the braided Platonic solids of Section 9.1 had their symmetry reduced because of the coloring on the faces. But we have seen in Section 9.7 that it is possible to braid the tetrahedron, octahedron, and icosahedron in such a way as to retain all the symmetry of the original polyhedron. Furthermore, in Sections 9.3 and 9.4 we obtained the diagonal cube and the golden dodecahedron, respectively. So we see that each of the Platonic solids can be braided in such a way as to preserve its underlying symmetry.

Meanwhile, returning to the discussion of symmetry, we note that the notion "more symmetric" above is really too restrictive. For we would like to be able to say that the regular n-gon becomes more symmetric as n increases. We are thus led to a weaker notion which will be useful provided we are dealing with figures with finite symmetry groups (e.g. polygons and polyhedra). We could then say – and do say – that A is more symmetric than B if G_A has more elements than G_B. Thus we have, in fact, two notions whereby we may compare symmetry – and they have the merit of being consistent. Indeed, if A is more symmetric than B in the first sense, it is more symmetric than B in the second sense – but not conversely.

---

[10] See Section 16.5 of this chapter for a careful treatment of even and odd permutations.
[11] This should be very familiar to those who have constructed Jennifer's puzzle of Chapter 8.

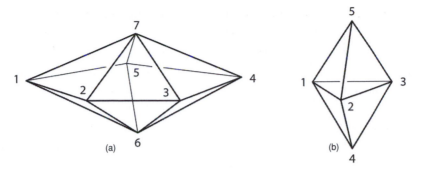

Figure 16.2 (a) Pentagonal dipyramid with labeled vertices. (b) Triangular dipyramid with labeled vertices.

Notice that we deliberately avoid the statement – often to be found in popular writing – "A is a symmetric figure." We regard this statement as having no precise meaning!

*16.3 Homologues

George Pólya, who made great contributions not only to mathematics itself, but also to the understanding of how and why we do mathematics – or perhaps one should say, "how and why we should do mathematics" – was particularly fascinated by the Platonic solids and first introduced his notion of *homologues* (orally) in connection with the study of their symmetry; we know that the idea of homologues played a key role in his thinking about one of his greatest contributions to the branch of mathematics known as *combinatorics*, namely, the ***Pólya enumeration theorem*** (see [68] for an intuitive account and [52] for a detailed account). Let us describe this notion of homologues in terms of symmetry groups. We believe that we are thereby increasing the scope of the notion while entirely maintaining the spirit of Pólya's original idea.

Let A be a geometrical configuration in a geometry given by a group Γ, and let the symmetry group of A with respect to this geometry be Γ_A; and let B be a subset of A. Thus, for example, A may be a polyhedron and B a face of that polyhedron. We consider the subgroup Γ_{AB} of Γ_A consisting of those motions in the symmetry group Γ_A of A which map B to itself. We consider a right coset of Γ_{AB} in Γ_A, that is, a set $\Gamma_{AB}g$, $g \in \Gamma_A$. Every element in $\Gamma_{AB}g$ sends B to the same subset Bg of A. The collection of these subsets is what Pólya called the collection of ***homologues*** of B in A. We see that the set of homologues of B in A is in one-one correspondence with the set of right cosets of Γ_{AB} in Γ_A.

Example 16.4 Consider the pentagonal dipyramid A of Figure 16.2(a). We may specify any motion in the symmetry group of A by the resulting permutation of its vertices 1, 2, 3, 4, 5, 6, 7; note that the polar vertices are 6 and 7. In fact, Γ_A is the dihedral group D_5, with 10 elements, given by the following permutations

$$(1\,2\,3\,4\,5\,6\,7) \begin{cases} \to (1\,2\,3\,4\,5\,6\,7) & \text{(identity)} \\ \to (2\,3\,4\,5\,1\,6\,7) & \text{(rotation through } \frac{2\pi}{5} \text{ about axis 67)} \\ \to (3\,4\,5\,1\,2\,6\,7) & \\ \to (4\,5\,1\,2\,3\,6\,7) & \\ \to (5\,1\,2\,3\,4\,6\,7) & \\ \to (5\,4\,3\,2\,1\,7\,6) & \text{(interchanging the poles)} \\ \to (4\,3\,2\,1\,5\,7\,6) & \text{(interchange plus rotation through } \frac{2\pi}{5}) \\ \to (3\,2\,1\,5\,4\,7\,6) & \\ \to (2\,1\,5\,4\,3\,7\,6) & \\ \to (1\,5\,4\,3\,2\,7\,6) & \end{cases}$$

First, let B be the edge 16. Then $\Gamma_{AB} = \{\text{Id}\}$, since only the identity sends the subset $(1, 6)$ to itself. Thus the index of Γ_{AB} in Γ_A is 10, and there are 10 homologues of the edge 16; these are the 10 "spines" of the dipyramid (i.e., we exclude the edges around the equator). Second, let B be the edge 12. Then Γ_{AB} has 2 elements, since there are 2 elements of Γ_A, namely the identity and $(1\,2\,3\,4\,5\,6\,7) \longrightarrow (2\,1\,5\,4\,3\,7\,6)$, which send the subset $(1, 2)$ to itself. Thus the index of Γ_{AB} in Γ_A is 5, and there are 5 homologues of the edge 12; these are the 5 edges around the equator.

Third, let B be the face $(1\,2\,6)$. Then $\Gamma_{AB} = \{\text{Id}\}$, so that, as in the first case, there are 10 homologues of the face $(1\,2\,6)$. In other words all the (triangular) faces are homologues of each other.

It is an easy, and informative, exercise to verify the following:

(1) Consider the triangular dipyramid P of Figure 16.2(b). Specify the motions of the symmetry group by writing down the resulting permutations of its vertices 1, 2, 3, 4, 5; note that the polar vertices are 4 and 5. In fact, show that Γ_P is the dihedral group D_3, with 6 elements.

(2) Check that the number of homologues for the edge 12 is 3; the number of homologues for the vertex 4 is 2; and the number of homologues for the face 124 is 6.

Let us now explain the Pólya enumeration theorem – actually, there are *two* theorems – and see how the notion of homologue fits into the story.

*16.4 The Pólya enumeration theorem

Let X be a finite set; the reader might like to keep in mind the set of vertices (or edges, or faces) of a polygon or polyhedron; and let G be a finite symmetry group acting on X. Suppose X has n elements, and that G has m elements; we write $|X| = n$, $|G| = m$. We may represent the elements of the set X by the integers

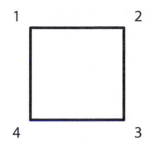

Figure 16.3   We denote the labeling of the vertices of the square by $\frac{1}{4}\blacksquare\frac{2}{3}$.

$1, 2, \ldots, n$. If $g \in G$, then g acts as a *permutation* of $\{1, 2, \ldots, n\}$. Now every permutation is uniquely expressible as a composition of cyclic permutations on mutually exclusive subsets of the elements of X. For example, the permutation

$$\begin{pmatrix} 1 & 2 & 3 & 4 & 5 & 6 & 7 & 8 & 9 & 10 & 11 \\ 2 & 4 & 10 & 1 & 3 & 11 & 7 & 6 & 8 & 5 & 9 \end{pmatrix} \tag{16.4}$$

of the set X of integers $\{1, 2, \ldots, 11\}$ is the composite

$$(1\ 2\ 4)(3\ 10\ 5)(6\ 11\ 9\ 8)(7),$$

where, e.g. $(1\ 2\ 4)$ denotes the cyclic permutation $\begin{pmatrix} 1 & 2 & 4 \\ 2 & 4 & 1 \end{pmatrix}$.

Thus the permutation (16.4) is the composite of one cyclic permutation of length 1, two cyclic permutations of length 3, and one cyclic permutation of length 4, the cyclic permutations acting on disjoint subsets of the set X. In general a permutation of X has the *type* $\{a_1, a_2, \ldots, a_n\}$ if it consists of a_1 permutations of length 1, a_2 permutations of length 2, $\ldots$, a_n permutations of length n, the permutations having disjoint domains of action; notice that $\sum_{i=1}^{n} i a_i = n$. For example, the permutation (16.4) has the type $\{1, 0, 2, 1, 0, 0, 0, 0, 0, 0, 0\}$. If g has the type $\{a_1, a_2, \ldots, a_n\}$, we define the *cycle index* of g to be the monomial

$$Z(g) = Z(g; x_1, x_2, \ldots, x_n) = x_1^{a_1} x_2^{a_2} \cdots x_n^{a_n}.$$

The *cycle index of G* is $Z(G) = Z(G; x_1, x_2, \ldots, x_n) = \frac{1}{m} \sum_{g \in G} Z(g)$.

We give an example which we will revisit periodically throughout this section.

Example 16.5 We consider the symmetries[12] of a square as shown in Figure 16.3.

The group G of symmetries is the group D_4 of order 8, which we here describe as a group of permutations of the set of vertices $\{1, 2, 3, 4\}$. Thus $n = 4, m = 8$, and the elements of G are

---

[12] Recall that, in considering the symmetries of a polygon, we permit reflections in 3-dimensional space about a line.

g_1(identity)   1 2 3 4 $\longrightarrow$ 1 2 3 4   cycle index x_1^4

g_2 1 2 3 4 $\longrightarrow$ 2 3 4 1   cycle index x_4

g_3 1 2 3 4 $\longrightarrow$ 3 4 1 2   cycle index x_2^2

g_4 1 2 3 4 $\longrightarrow$ 4 1 2 3   cycle index x_4

g_5 1 2 3 4 $\longrightarrow$ 3 2 1 4   cycle index $x_1^2 x_2$

g_6 1 2 3 4 $\longrightarrow$ 1 4 3 2   cycle index $x_1^2 x_2$

g_7 1 2 3 4 $\longrightarrow$ 2 1 4 3   cycle index x_2^2

g_8 1 2 3 4 $\longrightarrow$ 4 3 2 1   cycle index x_2^2.

Thus the cycle index of G is $\frac{1}{8}(x_1^4 + 2x_1^2 x_2 + 3x_2^2 + 2x_4)$.

Now suppose we want to *color* the elements of X; that is, we have a finite set Y of colors, $|Y| = r$, and a *coloring* of X is a function[13] $f : X \to Y$. For any $g \in G$, we regard the colorings f and fg as indistinguishable or *equivalent*; and a *pattern* is an equivalence class of colorings. Then Pólya's first theorem is as follows.

The cycle index theorem *The number of patterns is $Z(G; r, r, \ldots, r)$.*

Example 16.5 (continued) Suppose the vertices are to be colored red or blue. Then $r = 2$, and the number of patterns is $\frac{1}{8}(16 + 16 + 12 + 4) = 6$. In fact, the patterns are represented by the 6 colorings

$\begin{smallmatrix} R & R \\ & \blacksquare & \\ R & R \end{smallmatrix}$ $\begin{smallmatrix} B & R \\ & \blacksquare & \\ R & R \end{smallmatrix}$ $\begin{smallmatrix} B & B \\ & \blacksquare & \\ R & R \end{smallmatrix}$ $\begin{smallmatrix} B & R \\ & \blacksquare & \\ R & B \end{smallmatrix}$ $\begin{smallmatrix} R & B \\ & \blacksquare & \\ B & B \end{smallmatrix}$ $\begin{smallmatrix} B & B \\ & \blacksquare & \\ B & B \end{smallmatrix}$

Notice that we regard the colorings $\begin{smallmatrix} B & R \\ \blacksquare \\ R & R \end{smallmatrix}$ and $\begin{smallmatrix} R & B \\ \blacksquare \\ R & R \end{smallmatrix}$, for example, as indistinguishable or equivalent; we are sure you see why.

We now describe Pólya's second theorem. This is really the big theorem and the first theorem is, in fact, deducible from it. Let us enumerate the elements of Y (the colors) as $y_1, y_2, \ldots, y_r$.

The Pólya enumeration theorem *Evaluate the cycle index*

$$Z(G; x_1, x_2, \ldots, x_n) \quad at \; x_i = \sum_{j=1}^{r} y_j^i, i = 1, 2, \ldots, n.$$

Then the coefficient of $y_1^{n_1} y_2^{n_2} \cdots y_r^{n_r}$ is the number of patterns assigning the color y_j to n_j elements[14] of X.

Example 16.5 (continued) For the symmetries of the square we know that

$$Z(G) = \frac{1}{8}(x_1^4 + 2x_1^2 x_2 + 3x_2^2 + 2x_4).$$

---

[13] We speak of a *coloring* of X; this may be literally true, or it may merely be a metaphor for a rule for dividing the elements of X into disjoint classes.

[14] Of course $\sum_{j=1}^{r} n_j = n$.

Let $Y = \{R, B\}$; then the evaluation of $Z(G)$ at $x_i = R^i + B^i$, $i = 1, 2, 3, 4$, yields

$$\frac{1}{8}\left((R + B)^4 + 2(R + B)^2(R^2 + B^2) + 3(R^2 + B^2)^2 + 2(R^4 + B^4)\right)$$

$$= R^4 + R^3 B + 2R^2 B^2 + R B^3 + B^4.$$

(It is, of course, no coincidence that this polynomial is homogeneous (of degree $|X|$) and symmetric; and has integer coefficients.)

Thus the Pólya enumeration theorem tells us that there is 1 pattern with 4 red vertices (obvious); 1 pattern with 3 red vertices and 1 blue vertex (represented by the coloring $\genfrac{}{}{0pt}{}{B}{R}\blacksquare\genfrac{}{}{0pt}{}{R}{R}$); 2 patterns with 2 red vertices and 2 blue vertices (represented by the colorings $\genfrac{}{}{0pt}{}{B}{R}\blacksquare\genfrac{}{}{0pt}{}{B}{R}$ and $\genfrac{}{}{0pt}{}{B}{R}\blacksquare\genfrac{}{}{0pt}{}{R}{B}$); and the remaining possibilities are most easily analyzed by considerations of symmetry.

Consider the various coloring functions $X \to Y$ representing a given pattern. These functions all have the form $fg : X \to Y$, where f is a fixed coloring and g ranges over the elements of G. It will turn out that the colorings fg are essentially the **homologues of f**. Let us first revert to our example.

Example 16.5 (continued) As we have seen, there is one coloring in which all vertices are colored red. There is only one homologue, namely, $\genfrac{}{}{0pt}{}{R}{R}\blacksquare\genfrac{}{}{0pt}{}{R}{R}$.

There is one coloring in which 3 vertices are colored red and 1 blue. There are 4 homologues, namely,

$$\genfrac{}{}{0pt}{}{B}{R}\blacksquare\genfrac{}{}{0pt}{}{R}{R} \quad \genfrac{}{}{0pt}{}{R}{R}\blacksquare\genfrac{}{}{0pt}{}{B}{R} \quad \genfrac{}{}{0pt}{}{R}{R}\blacksquare\genfrac{}{}{0pt}{}{R}{B} \quad \genfrac{}{}{0pt}{}{R}{B}\blacksquare\genfrac{}{}{0pt}{}{R}{R}$$

There are 2 colorings in which 2 vertices are colored red and 2 blue. In the first there are 4 homologues, namely,

$$\genfrac{}{}{0pt}{}{B}{R}\blacksquare\genfrac{}{}{0pt}{}{B}{R} \quad \genfrac{}{}{0pt}{}{R}{R}\blacksquare\genfrac{}{}{0pt}{}{B}{B} \quad \genfrac{}{}{0pt}{}{R}{B}\blacksquare\genfrac{}{}{0pt}{}{R}{B} \quad \genfrac{}{}{0pt}{}{B}{B}\blacksquare\genfrac{}{}{0pt}{}{R}{R}$$

In the second there are 2 homologues, namely,

$$\genfrac{}{}{0pt}{}{B}{R}\blacksquare\genfrac{}{}{0pt}{}{R}{B} \quad \genfrac{}{}{0pt}{}{R}{B}\blacksquare\genfrac{}{}{0pt}{}{B}{R}$$

The analysis may be completed by considerations of symmetry.

Let us now show how this concept of homologues agrees with our earlier definition. We are given the group G of permutations[15] of X. Given a coloring $f : X \to Y$, we consider the subset G_0 of G consisting of those g such that $fg = f$, that is, those movements of X which preserve the coloring. It is easy to see (just as easy as in our earlier, simpler situation) that G_0 is a *subgroup* of G. Corresponding to each coset $G_0 g$ of G_0 in G we have a coloring fg of X, and

---

[15] Recall, from Section 16.3, the group Γ_A of symmetries of the configuration A.

Table 16.1.

Description of rotation as an axis through the center of	the amount of rotation	the number of rotations of this type
Opposite faces	$\pm\frac{1}{4}$ turn	6
Opposite faces	$\frac{1}{2}$ turn	3
Opposite vertices	$\pm\frac{1}{3}$ turn	8
Opposite edges	$\frac{1}{2}$ turn	6
Identity	0	1
		Total number of rotations $= 24$

Table 16.2.

Description of the type of permutation	the number of permutations of this type
$(\_\_\_\_) \left(\begin{smallmatrix}\text{one cycle of} \\ \text{length } 4\end{smallmatrix}\right)$	6
$(\_\_)(\_\_) \left(\begin{smallmatrix}\text{two cycles, each} \\ \text{of length } 2\end{smallmatrix}\right)$	3
$(\_\_\_)(\_) \left(\begin{smallmatrix}\text{one cycle of length } 3 \\ \text{and one cycle of length } 1\end{smallmatrix}\right)$	8
$(\_\_)(\_)(\_) \left(\begin{smallmatrix}\text{one cycle of length } 2 \\ \text{and two cycles of length } 1\end{smallmatrix}\right)$	6
$(\_)(\_)(\_)(\_) \left(\begin{smallmatrix}\text{four cycles of length } 1, \\ \text{i.e. the identity}\end{smallmatrix}\right)$	1
	Total number of permutations $= 24$

these colorings run through the pattern determined by f. We describe the set of colorings $\{fg\}$ as the set of *homologues* of the coloring f; just as in our geometric definition in Section 16.3, they are in one-one correspondence with the set of right cosets of G_0 in G.

As an example of homologues, take the diagonal cube described in Section 9.3 and use it to verify the entries in Table 16.1.

Compare this table with Table 16.2 concerning the permutations of 4 objects.

Notice anything? (**Hint:** Think of the strips of your cube as being numbered 1, 2, 3, 4 and observe what happens to the strips when you perform the rotations in Table 16.1. This is one, very vivid, way to see why the symmetry group of the cube is S_4. Of course, there are other ways, too. For example, we saw in Section 10.1 how the rotations of the cube simply permute the 4 interior diagonals of the cube. It is certainly harder to "see" the interior diagonals than it is to see the strips of the diagonal cube.

*16.5 Even and odd permutations

We described the symmetry group of a regular tetrahedron as the alternating group A_4 consisting of *even* permutations of the set (1, 2, 3, 4). We owe you at the very least a definition of this concept! The reason we postponed giving you this is that the *precise* definition is difficult to understand. See if you agree with us.

Let A be the *alternating form in n variables*, that is, the expression

$$A = \prod_{1 \le i < j \le n} (x_i - x_j). \tag{16.5}$$

Then any permutation ρ in S_n acts on A by permuting the suffixes $i, j, \ldots$ appearing in (16.5) by means of ρ; thus

$$A\rho = \prod_{1 \le i < j \le n} (x_{i\rho} - x_{j\rho}). \tag{16.6}$$

We claim that $A\rho$ is either A or $-A$. For, given any factor $x_i - x_j$ of $A\rho$, then $i = h\rho$, $j = k\rho$, for some (unique) h, k between 1 and n; and either $x_h - x_k$ or $x_k - x_h$ occurred in A. Thus we are led to the crucial definition:

Definition of even and odd permutations

The permutation ρ is *even* if $A\rho = A$; it is *odd* if $A\rho = -A$. We call the evenness or oddness of a permutation its **parity**.

Example 16.6 Consider the *cyclic* permutation (1 2 3), that is, $\rho(1) = 2$, $\rho(2) = 3$, $\rho(3) = 1$. Then if $A = (x_1 - x_2)(x_1 - x_3)(x_2 - x_3)$,

$$A\rho = (x_2 - x_3)(x_2 - x_1)(x_3 - x_1) = A.$$

Thus ρ is even. On the other hand, if σ is the permutation (1 3), that is, $\sigma(1) = 3$, $\sigma(2) = 2$, $\sigma(3) = 1$, then $A\sigma = (x_3 - x_2)(x_3 - x_1)(x_2 - x_1) = -A$, so σ is odd.

Now in the next section you will see that every permutation may be expressed as a composition of *transpositions*, that is, of cycles which (like (1 3) above) merely interchange two numbers. Granted this, one may show the following:

The even permutation theorem *A permutation is even if and only if it may be expressed as a composition of an even number of transpositions.*

You may ask – why don't we simply *define* an even permutation as one which may be expressed as a composition of an even number of transpositions? For that is surely the reason for the use of the name "even." The answer is that it is by no means obvious that, if a permutation can be expressed as a composition of an *even* number of transpositions, it cannot also be expressed as a composition of an *odd* number of transpositions. It is, indeed, the even permutation theorem which makes that fact clear.

However, before we prove the even permutation theorem, we'll take a break and show you why every permutation may be expressed as a composition of transpositions.[16]

We need a few steps.

Step 1

Satisfy yourself, by looking at a particular but not special case, that every permutation of a set of numbers may be expressed as a composition of cycles acting on disjoint sets of numbers. (See (16.4).)

Step 2

We now need to show that every cycle $(\iota_1, \iota_2, \ldots, \iota_r)$ is a composition of transpositions. Of course this holds if $r = 2$. Now, if $r \geq 3$, we show that

$$(\iota_1, \iota_2, \ldots, \iota_r) = (\iota_1, \iota_2, \ldots, \iota_{r-1})(\iota_1, \iota_r)$$

and hence deduce, by induction on r, that the cycle $(\iota_1, \iota_2, \ldots, \iota_r)$ is a composition of transpositions.

Step 3

Find an explicit expression for $(\iota_1, \iota_2, \ldots, \iota_r)$ as a composition of transpositions.

Now we return to the proof of the even permutation theorem. It is obvious that even and odd permutations, under composition $*$, act like ordinary integers under addition, that is,

$$\text{even} * \text{even} = \text{even},$$
$$\text{odd} * \text{odd} = \text{even},$$
$$\text{even} * \text{odd} = \text{odd},$$
$$\text{odd} * \text{even} = \text{odd}.$$

It therefore follows immediately that, if we can show that every transposition is odd, then it must be true that any permutation expressible as a composition of an *even* number of transpositions is *even*, and any permutation expressible as a composition of an *odd* number of transpositions is *odd* – and this will prove the

---

[16] We adopt the usual convention that the identity permutation is the *empty* composition of transpositions.

even permutation theorem, Thus we are left to prove that *every transposition is odd.*

First, we claim that the transposition (1 2) is odd. For if we apply (1 2) to the form $A = \prod_{1 \leq i < j \leq n}(x_i - x_j)$, then the factor $(x_1 - x_2)$ is transformed into its negative, while all the other factors are merely permuted among themselves with no change of sign. Thus (1 2) is an odd permutation. Second, consider the transposition $(i \ j)$, $i < j$. We claim that

$$(2 \ j)(1 \ 2)(2 \ j) = (1 \ j), \ j \geq 3,$$

$$(1 \ j)(1 \ 2)(1 \ j) = (2 \ j), \ j \geq 3,$$

$$(1 \ i)(2 \ j)(1 \ 2)(1 \ i)(2 \ j) = (i \ j).$$

Thus, since (1 2) is odd, so is $(i \ j)$, and the even permutation theorem is completely proved.[17]

Now it is plain that the set of even permutations of n objects is a *subgroup* of S_n; it is called the **alternating group on n objects** and written A_n. Moreover, if $n \geq 2$ and ρ is any *odd* permutation then $A_n\rho$ is the set of *all* odd permutations. For certainly every permutation in $A_n\rho$ is odd; and conversely, if σ is odd, then $\sigma\rho^{-1}$ is even, $\sigma\rho^{-1} \in A_n$, and $\sigma = (\sigma\rho^{-1})\rho$. Thus S_n is the disjoint union of the *two* cosets A_n and $A_n\rho$, consisting of the even and odd permutations respectively, and hence A_n is a subgroup of S_n of index 2, as claimed in Section 16.2.

Notice that we could have argued above using ρA_n instead of $A_n\rho$; indeed, and remarkably, $\rho A_n = A_n\rho$. It is a special, and very important, property of some subgroups H of a group G that $gH = Hg$ for all $g \in G$. We call such subgroups **normal**. Our argument above can easily be generalized to show that every subgroup of index 2 of an arbitrary group G is normal.

Finally, let us remark here that the reason that even permutations figure so prominently in the discussion of symmetry in geometry is this: any symmetry of a polyhedron, and any orientation-preserving symmetry of a polygon, will induce an *even* permutation of the vertices. Try it and see!

It is a fact that the regular tetrahedron (T) may be inscribed in the regular hexahedron (H, or cube), as shown in Figure 16.4. Look at the rotations of the cube that leave T occupying its original position within the cube. You should see that these rotations are simply the *even* permutations of the 4 strips of the diagonal cube (and that all the odd permutations move the tetrahedron from its original position to the same new position). Thus you have a dramatic confirmation that the rotation group of T is A_4, the alternating subgroup of S_4, the group of rotations of H.

---

[17] Notice that the parity of a permutation is unaffected by the choice of n in the definition of even and odd permutations.

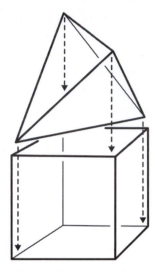

Figure 16.4   How to fit a tetrahedron inside a cube.

Figure 16.5 was copied by JP from Pólya's personal notebook. See if you can fill in the blanks.

16.6 Epilogue: Pólya and ourselves – Mathematics, tea, and cakes

George Pólya (1887–1985) emigrated to the United States from Europe in 1940 and joined the Mathematics Department at Stanford University in 1942. Although the rest of his professional life was spent at Stanford, he made many trips abroad to accept visiting appointments for short periods of time. During Pólya's visit in 1966 to the famous Eidgenössische Technische Hochschule in Zürich (always abbreviated to ETH) he shared an office with Peter Hilton (and PH was a guest at his 80th birthday party, held in Zürich, in 1967).

In 1969 Pólya was invited by Gerald L. Alexanderson (then Mathematics Department Chairman at Santa Clara University) to give a colloquium talk at SCU. While there, GP met JP, also on the mathematics faculty at SCU, and was fascinated (a) by the models in her office (almost all of which are described in this book) and (b) by her lack of knowledge about their symmetry and their usefulness in exemplifying some of the mathematics of polyhedral geometry. After this initial meeting, intrigued by her remarkable combination of properties (a) and (b) above, GP invited JP regularly once a week to visit him and his wife Stella at their Palo Alto home; these visits continued until his death in 1985.[18] The Pólyas were welcome guests at the Pedersen home for Thanksgiving dinner for many successive years.

---

[18] After George Pólya's death, JP continued to visit Stella Pólya at least once a week until her death in 1989, just before her 94th birthday.

		T	H	O	D	I
F		4	6	8	12	20
V		4	8	6	20	12
E		6	12	12	30	30
Axes of sym-metry	$\frac{2\pi}{2}$	?	6			?
	$\frac{2\pi}{3}$	?	4			?
	$\frac{2\pi}{4}$		3			
	$\frac{2\pi}{5}$					?
Planes of symmetry			6			?
			3			
Faces angles		?	24			?
Diagonals (interior)			4	3	60	?
					30	?
					10	

Figure 16.5   Here T is tetrahedron, H is dodecahedron (or cube), etc.

A typical visit by JP to the Pólya's included a discussion with GP about mathematics. After an hour or so Stella would appear with tea and cakes, or cookies, and the mathematical conversation would give place to a discussion of current events, politics, and other interesting matters. It was during these pre-refreshment periods that JP learned about rotation groups (knowledge that GP acquired from the master Felix Klein himself) and the Pólya enumeration theorem, about Euler's famous formula connecting the vertices, edges, and faces of a polyhedron, and about the formula Descartes discovered concerning the total angular deficiency of a polyhedron. JP found herself studying very hard and looking forward each week to discussing the new-found aspects of her own models. GP and JP also discussed pedagogy and, in fact, JP was GP's last co-author (see [65]).

In 1978 JP was asked to try to get GP and PH together[19] in Seattle at the joint annual meeting of the American Mathematical Society and the Mathematical Association of America, to discuss "How to and How *Not* to Teach Mathematics."

---

[19] This was how PH and JP met and began a collaboration that has resulted in over 100 papers and 5 books – so far!

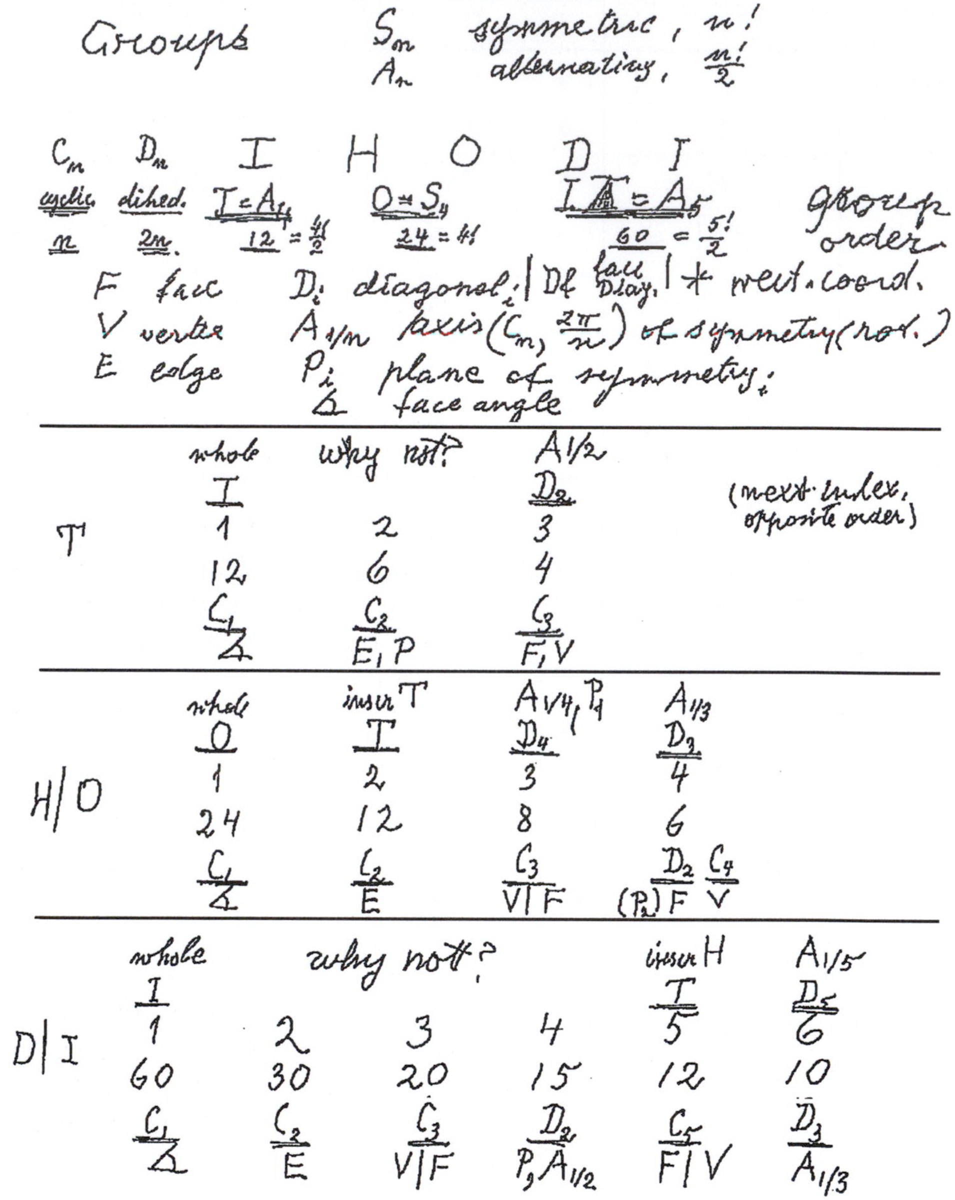

Groups S_n symmetric, $n!$
A_n alternating, $\frac{n!}{2}$

C_n D_n I H O D I

cyclic   dihed. $I = A_4$ $O = S_4$ $I = A_5$ group
   n $2n$ $12 = \frac{4!}{2}$ $24 = 4!$ $60 = \frac{5!}{2}$ order

F face D_i diagonal. $|$ Df $\frac{face}{Diag.}$ $|$ $*$ rect. coord.
V vertex $A_{i/n}$ axis $(C_n, \frac{2\pi}{n})$ of symmetry (rot.)
E edge P_i plane of symmetry.
     $\angle$ face angle

	whole	why not?	$A_{1/2}$		
	I		D_2		(next index, opposite order)
T	1	2	3		
	12	6	4		
	$\frac{C_4}{\angle}$	$\frac{C_2}{E, P}$	$\frac{C_3}{F, V}$		

	whole	insert T	$A_{1/4}, P_i$	$A_{1/3}$
	O	I	D_4	D_3
H/O	1	2	3	4
	24	12	8	6
	$\frac{C_4}{\angle}$	$\frac{C_2}{E}$	$\frac{C_3}{V|F}$	$\frac{D_2 \, C_4}{(P_i) F \; V}$

	whole	why not?			insert H	$A_{1/5}$
	I				I	D_5
D/I	1	2	3	4	5	6
	60	30	20	15	12	10
	$\frac{C_1}{\angle}$	$\frac{C_2}{E}$	$\frac{C_3}{V|F}$	$\frac{D_2}{P, A_{1/2}}$	$\frac{C_5}{F|V}$	$\frac{D_3}{A_{1/3}}$

Figure 16.6   As before *T* is tetrahedron, *H* is hexahedron (or cube), etc.

The suggestion was that PH should discuss "How Not to Teach Mathematics" and this would be followed by GP giving "Some Rules of Thumb for Good Teaching." GP agreed to participate on the condition that JP would handle the travel details of getting him to and from Seattle. PH also gave only conditional approval for the plan. His hesitation sprang from his conviction that it would be much more interesting, and effective, if he (PH) were to *demonstrate* a thoroughly bad mathematics lecture (instead of simply talking about it). He also suggested that JP should be the moderator for the program.

All conditions were met and the Seattle presentation duly took place. It was a tremendous success. PH's part was hilarious and some said it nearly ruined the rest of the meeting as participants saw many of PH's intentional errors *unintentionally* repeated by some of the other speakers. GP's contribution was, as you might expect, superb and had the unmistakable mark of a master teacher. By request the session was repeated at the National Council of Teachers of Mathematics San Diego meeting in the fall of 1978.

As a result of the NCTM meeting PH visited SCU to give a colloquium talk and when he saw the models in JP's office they again sparked long discussions, but this time the discussions centered on the differences between the ways geometers and topologists classify surfaces.

In 1982, while PH was on sabbatical leave as a visiting professor at the ETH in Zürich and JP was visiting there for a quarter, they began looking seriously at the questions raised by paper-folding. (See [23, 24] and the present book for more details on the results of this study.) It was during this period that the FAT algorithm was born. This innocent-looking algorithm, in fact, opened the flood gates for the development both of the general folding procedures and of the number theory that grew out of the paper-folding – and which is featured in this book.

After 1978 whenever PH visited SCU he went with JP to visit the Pólyas and together they continued the tradition of mathematics, tea, and cakes. In 1981, PH and JP, along with Alexanderson, cooperated with GP to bring out the Combined Edition of *Mathematical Discovery* (see [70]).[20]

During many of the tea parties at the Pólya's home over the years, GP talked about his idea of homologues, and on one occasion told PH and JP that he had never written about them and that someday he would like us to write about them – in fact, he extracted a promise from us that we would do so. We are very happy to be able, in this chapter, to fulfill our promise to our dear friend and teacher George Pólya, and, at the same time, to convey the flavor, and a few of the details, of our friendly relationship with that remarkable man.

---

[20] Alexanderson updated the references, PH wrote a foreword, and JP provided an expanded (and less esoteric) index.

17

Returning to the number-theory thread – Generalized quasi-order and coach theorems

17.1 Setting the stage

In Chapter 7 we explicitly stated, and proved, the quasi-order theorem in base 2 and stated it in base 3, giving examples. We will carry this thread further in Section 17.2 and show how, in base 2, it leads to what we call the coach theorem. Having come this far it is natural to ask if the quasi-order theorem and the coach theorem can be generalized. The answer is "yes" and we carry out the details in Sections 17.3 and 17.4.

Sections 17.5 and 17.6 grew out of the interest shown by two students, Victor Quintanar-Zilinskas and Linda Velarde, in work done by the authors and Byron Walden [46]. After taking a course from one of the authors (JP), Victor and Linda asked if they could continue "looking for patterns" in the complete coaches in base t (as explained below), during a summer research project at Santa Clara University. Permission was readily granted, and Sections 17.5 and 17.6 chronicle their discoveries (which were published in [44]).

Although it is important to be able to create the complete symbols for oneself, it greatly facilitates looking for patterns to have a large amount of data available. So, at the beginning of the summer research project, Byron Walden wrote a Maple program that generates the desired data. Those of you interested in studying the complete symbols may wish to write your own program to obtain such data.

We leave the reader with some open questions in Section 17.7.

17.2 The coach theorem

In this section we will look at the quasi-order theorem again and, in particular, study some of the patterns in the complete symbols. For convenience let us restate the quasi-order theorem (that first appeared in Chapter 7).

The quasi-order theorem *Suppose that a, b are odd, with $a = a_1 < \frac{b}{2}$, and the symbol*

$$b \begin{vmatrix} a_1 & a_2 & \cdot & \cdot & \cdot & a_r \\ k_1 & k_2 & \cdot & \cdot & \cdot & k_r \end{vmatrix} \tag{17.1}$$

is obtained using the calculation

$$b - a_i = 2^{k_i} a_{i+1} \quad \text{(with } k_i \text{ maximal)}. \tag{17.2}$$

*Let $\sum_{i=1}^r k_i = k$, and assume that (17.1) is not only contracted ($a_{r+1} = a_1$), but also reduced ($\gcd(a_i, b) = 1$). Then the quasi-order of 2 mod b is k. That is, k is the **smallest** positive integer such that*

$$2^k \equiv \pm 1 \mod b.$$

In fact, $2^k \equiv (-1)^r \mod b$. (This means that b exactly divides $2^k - (-1)^r$, which may be written as $b \mid 2^k - (-1)^r$.)

Example 17.1 Consider combining all the possible symbols for $b = 31$ into one *complete* symbol, which we will write $\Sigma(31)$, calling each part of $\Sigma(31)$ a *coach* (because it looks like a coach on a train), with the number of coaches being denoted by c.

The complete symbol $\Sigma(31)$ then takes the form

$$\Sigma(31): \quad 31 \begin{vmatrix} 1 & 15 & 3 & 7 & 5 & 13 & 9 & 11 \\ 1 & 4 & 2 & 3 & 1 & 1 & 1 & 2 \end{vmatrix}. \tag{17.3}$$

What do you notice about the value of the bottom row sum in each coach of (17.3)? It is the same, namely 5. Notice, too, that the parity in the number of entries is the same in each coach (namely, in this case, even). Is this an accident? The interested reader should try writing out a few complete $\Sigma(b)$ symbols for odd numbers b of your choice and look for patterns among the numbers involved. The quasi-order theorem applies to *any* coach of a complete $\Sigma(b)$ symbol.

Example 17.1 also gives us a glimpse of what is to come. Notice that $\Phi(31) = 30$, where Φ is the Euler totient function,[1] $c = 3$, and $k = 5$, so that, in this case $\Phi(b) = 2ck$. Here are a couple of other complete symbols for you to examine

---

[1] The Euler totient function, Φ, counts the number of positive numbers less than a positive number b that are relatively prime to b. It is a well-known result of number theory that if p is a prime then $\Phi(p^n) = (p-1)p^{n-1}$. Furthermore, for every pair of mutually prime positive numbers k and ℓ, $\Phi(k \cdot \ell) = \Phi(k)\Phi(\ell)$. Thus we can calculate $\Phi(m)$ for any positive integer m. Another, equivalent, definition of Φ is

$$\Phi(n) = n \left(1 - \frac{1}{p_1} \right) \left(1 - \frac{1}{p_2} \right) \cdots \left(1 - \frac{1}{p_k} \right),$$

where $p_1, p_2, \ldots, p_k$ are the distinct prime factors of n.

to see whether the relationships you have observed are just happy accidents, or whether you believe that they must always happen. (Of course, you may construct some for yourself as well, just to make sure we haven't chosen the only ones for which the relation $\Phi(b) = 2ck$ holds.) Thus

$$\Sigma(43): \quad 43 \begin{vmatrix} 1 & 21 & 11 & 3 & 5 & 19 & 7 & 9 & 17 & 13 & 15 \\ 1 & 1 & 5 & 3 & 1 & 3 & 2 & 1 & 1 & 1 & 2 \end{vmatrix},$$
$$\Phi(43) = 42, \quad k = 7, \quad c = 3.$$

$$\Sigma(51): \quad 51 \begin{vmatrix} 1 & 25 & 13 & 19 & 5 & 23 & 7 & 11 \\ 1 & 1 & 1 & 5 & 1 & 2 & 2 & 3 \end{vmatrix},$$
$$\Phi(51) = 32, \quad k = 8, \quad c = 2.$$

$$\Sigma(65): \quad 65 \begin{vmatrix} 1 & 3 & 31 & 17 & 7 & 29 & 9 & 11 & 27 & 19 & 23 & 21 \\ 6 & 1 & 1 & 4 & 1 & 2 & 3 & 1 & 1 & 1 & 1 & 2 \end{vmatrix},$$
$$\Phi(65) = 48, \quad k = 6, \quad c = 4.$$

$$\Sigma(33): \quad 33 \begin{vmatrix} 1 & 5 & 7 & 13 \\ 5 & 2 & 1 & 2 \end{vmatrix},$$
$$\Phi(33) = 20, \quad k = 5, \quad c = 2.$$

Now we enunciate our theorem about $\Sigma(b)$.

The coach theorem *Let $b > 1$ be an odd number, and let $\Phi(b)$ be the Euler totient function of b. Form $\Sigma(b)$, the complete symbol of b, and let c be the number of* **coaches** *in $\Sigma(b)$ and $k = \sum_{i=1}^{r} k_i$. Then $\Phi(b) = 2ck$.*[2]

One can see what is going on by looking carefully at a particular, but not special, case. Thus, consider the complete symbol

$$\Sigma(33): \quad 33 \begin{vmatrix} 1 & 5 & 7 & 13 \\ 5 & 2 & 1 & 2 \end{vmatrix},$$
$$k = 5, \quad c = 2, \quad \text{and} \quad \Phi(33) = 20.$$

First, construct a **complete modified symbol**, corresponding with $\Sigma(33)$, as follows:[3]

$$33 \begin{pmatrix} 1 & & 1 \\ & 5 & \end{pmatrix} \begin{pmatrix} 13 & & 7 & & 5 & & 13 \\ & 1 & & 2 & & 2 & \end{pmatrix}. \tag{17.4}$$

---

[2] A generalization of this theorem in base t (thus not restricted to $t = 2$), and its proof, appear in [32].
[3] A modified symbol was defined in Section 7.2 for a single coach and was used there to prove the quasi-order theorem in base 2.

First notice that, from geometric considerations, a $\left\{\frac{33}{a}\right\}$-gon is also a $\left\{\frac{33}{33-a}\right\}$-gon (which we may think of as the **complementary** $\left\{\frac{33}{a}\right\}$-gon). Since it is true that, if $\gcd(a, 33) = 1$, it follows that $\gcd(33 - a, 33) = 1$, the complementary polygons are relevant to the number theory (though not to the geometry). Thus we list here the $\left\{\frac{33}{a}\right\}$-gons we get from the folding instructions for the first modified coach, along with the equivalent complementary $\left\{\frac{33}{33-a}\right\}$-gons in the following row. The 5 star polygons connected with the modified first coach of (17.4) are:

$$\left\{\tfrac{33}{1}\right\}\text{-}, \quad \left\{\tfrac{33}{2}\right\}\text{-}, \quad \left\{\tfrac{33}{4}\right\}\text{-}, \quad \left\{\tfrac{33}{8}\right\}\text{-} \quad \text{and} \quad \left\{\tfrac{33}{15}\right\}\text{-gons} \qquad (17.5)$$

and their 5 complementary star polygons are the

$$\left\{\tfrac{33}{32}\right\}\text{-}, \quad \left\{\tfrac{33}{31}\right\}\text{-}, \quad \left\{\tfrac{33}{29}\right\}\text{-}, \quad \left\{\tfrac{33}{25}\right\}\text{-} \quad \text{and} \quad \left\{\tfrac{33}{16}\right\}\text{-gons.} \qquad (17.5')$$

Similarly, the 5 star polygons connected with the second modified coach of (17.4) are:

$$\left\{\tfrac{33}{13}\right\}\text{-}, \quad \left\{\tfrac{33}{7}\right\}\text{-}, \quad \left\{\tfrac{33}{14}\right\}\text{-}, \quad \left\{\tfrac{33}{5}\right\}\text{-} \quad \text{and} \quad \left\{\tfrac{33}{10}\right\}\text{-gons} \qquad (17.6)$$

and their 5 complementary star polygons are the

$$\left\{\tfrac{33}{20}\right\}\text{-}, \quad \left\{\tfrac{33}{26}\right\}\text{-}, \quad \left\{\tfrac{33}{19}\right\}\text{-}, \quad \left\{\tfrac{33}{28}\right\}\text{-} \quad \text{and} \quad \left\{\tfrac{33}{23}\right\}\text{-gons.} \qquad (17.6')$$

Notice that for the $\left\{\frac{33}{a}\right\}$-gons listed in (17.5) and (17.6) the value of a will be either one of the a_i in the top row of (17.4) or it is of the form $a_i 2^p$, where $a_i 2^p < \frac{b}{2}$, with p a positive integer.

Since the proof of the quasi-order theorem was *independent* of the choice of a_i we know that the value of k must be the same for every coach. Furthermore, every possible $a_i < \frac{b}{2}$ must appear somewhere on the first row of some coach in the complete symbol, otherwise we would be compelled to construct another coach. Consequently, in this example, we have listed in (17.5) and (17.6) all possible star polygons (with their complementary polygons in (17.5') and (17.6')). This means we have accounted for *all* the numbers less than 33 that are relatively prime to 33. Note that the denominators in the list of star polygons appearing in (17.5') and (17.6') are, in each case, just the sequence of numbers that appear in item (17.12), of Section 7.2 (without the repeat of c_1), in the proof of the quasi-order theorem. Plainly, what worked above for $\Sigma(33)$ would work for any odd number b, so we have a "proof" of the coach theorem, which we don't believe would be any more convincing if we wrote it out in more traditional terms. Note that the 2 in the formula $\Phi(b) = 2ck$ is present because, as geometers, we count only the $\left\{\frac{b}{a_i}\right\}$-gons where $a_i < \frac{b}{2}$ and not their equivalent complementary $\left\{\frac{b}{b-a_i}\right\}$-gons.

The coach theorem admits the following two nice corollaries:

Corollary 17.1 *If b is odd and we form a symbol from b, and if k is the quasi-order of 2 mod b, then $k \mid \frac{1}{2}\Phi(b)$.*

Corollary 17.2 *If, for a given odd number b, it is only possible to construct one reduced, contracted symbol from b, then $\Phi(b) = 2k$, where k is the quasi-order of 2 mod b.*

The hypothesis in Corollary 16.2 is equivalent to saying that the complete symbol created by the odd number b has only one coach.

Of course, another proof of the coach theorem may be obtained, with mild variations, by letting $t = 2$ in Section 17.4 – but that is really using a hammer to crack a nut.

17.3 The generalized quasi-order theorem

How do we generalize the quasi-order theorem? It is interesting, and not altogether surprising, that our main difficulty in generalizing this theorem to a general base t lies not in *proving* the generalization but in *stating* it. For generalization is an art, not an algorithmic procedure, so judicious choices must be made in formulating the generalization. It is particularly striking that the appropriate generalization of the relation $a = a_1 < \frac{b}{2}$ is not, as we originally thought, $a = a_1 < \frac{b}{t}$, but $a = a_1 < \frac{b}{2}$! We are now ready to formalize the appropriate generalization.

The general quasi-order theorem *Suppose b, a are mutually prime, with b prime to the base t, where $t \nmid a$, and $a_1 = a < \frac{b}{2}$. Then construct the contracted $(a_{r+1} = a_1)$, reduced $(\gcd(a_i, b) = 1)$, t-symbol*

$$b \begin{vmatrix} a_1 & a_2 & \cdot & \cdot & \cdot & a_r \\ k_1 & k_2 & \cdot & \cdot & \cdot & k_r \\ \epsilon_1 & \epsilon_2 & \cdot & \cdot & \cdot & \epsilon_r \\ q_1 & q_2 & \cdot & \cdot & \cdot & q_r \end{vmatrix}_t \tag{17.7}$$

where

$$q_i b + (-1)^{\epsilon_i} a_i = t^{k_i} a_{i+1}, \quad i = 1, 2, \ldots, r \quad (a_{r+1} = a_1). \tag{17.8}$$

Moreover, when $\epsilon_i = 0$, we use $q_i b + a$, $1 \le q_i \le \frac{t}{2} - 1$,

and when $\epsilon_i = 1$, we use $q_i b - a$, $1 \le q_i \le \frac{t}{2}$,

choosing the smallest q_i such that $q_i + (-1)^{\epsilon_i}$ has t as a factor.

Then the quasi-order of t mod *b is* $k = \sum k_i$. *Indeed,*

$$t^k \equiv (-1)^E \bmod b,$$

where $E = \sum \epsilon_i$.

We call the algorithm that generates the symbol (17.4) from (17.5) the ψ-algorithm.

The proof of the general quasi-order theorem appears in [24] where the bottom row of the symbol, involving the q_i, doesn't appear because it isn't essential to the proof. We include the bottom row here because we have discovered that it plays a role when one is looking for patterns among the coaches of the complete symbol. This brings up the question: Does the coach theorem hold for any base t? Before we try to answer this, let us look at some examples of $\Sigma_t(b)$, where $\Sigma_t(b)$ is the complete symbol of b for the base t.

Example 17.2 Construct $\Sigma_4(17)$ and state what it means in terms of the quasi-order of 4 mod 17.

Solution For each coach we use the equation (beginning with the smallest available a_i)

$$qb + (-1)^{\epsilon_i} a_i = 4^{k_i} \cdot a_{i+1} \qquad \text{where} \quad q = 1 \text{ or } 2, \quad \epsilon_i = 0 \text{ or } 1.$$

A routine approach would be to calculate $b - a_i$, $b + a_i$, and $2b - a_i$ in that order, stopping at the first one that yields at least one factor of 4. Proceeding in this way and only recording the successive calculations, we obtain, for each coach, the calculations below, that are then recorded in the symbol (17.9).

First coach:	$17 - 1$	$= 4^2 \cdot 1$
Second coach:	$2 \cdot 17 - 2$	$= 4^2 \cdot 2$
Third coach:	$17 + 3$	$= 4 \cdot 5$
	$17 - 5$	$= 4 \cdot 3$
Fourth coach:	$2 \cdot 17 - 6$	$= 4 \cdot 7$
	$17 + 7$	$= 4 \cdot 6$

so that $\Sigma_4(17)$ is

$$17 \begin{vmatrix} 1 & 2 & 3 & 5 & 6 & 7 \\ 2 & 2 & 1 & 1 & 1 & 1 \\ 1 & 1 & 0 & 1 & 1 & 0 \\ 1 & 2 & 1 & 1 & 2 & 1 \end{vmatrix}_4, \qquad k = 2, \quad c = 4. \tag{17.9}$$

From any coach in (17.9) we see that the quasi-order of 4 mod 17 is 2. In fact,

$$4^2 \equiv (-1)^1 \bmod 17.$$

Furthermore, $\Phi(17) = 16 = 2 \cdot 2 \cdot 4$, where Φ is the Euler totient function (so Φ here satisfies the coach theorem).

Example 17.3 Construct $\Sigma_5(67)$ and state what it means in terms of the quasi-order of 5 mod 67.

Solution For each coach we use the equation (beginning with the smallest available a_i)

$$qb + (-1)^{\epsilon_i} a_i = 5^{k_i} \cdot a_{i+1} \qquad \text{where} \quad q = 1 \text{ or } 2, \quad \epsilon_i = 0 \text{ or } 1.$$

This time we calculate $b - a_i, b + a_i, 2b - a_i$, and $2b + a_i$ in that order, stopping at the first one that yields at least one factor of 5. The calculations for the first coach are

$$2 \cdot 67 + 1 = 5^1 \cdot 27$$
$$67 - 27 = 5^1 \cdot 8$$
$$67 + 8 = 5^2 \cdot 3$$
$$67 + 3 = 5^1 \cdot 14$$
$$2 \cdot 67 - 14 = 5^1 \cdot 24$$
$$2 \cdot 67 - 24 = 5^1 \cdot 22$$
$$67 - 22 = 5^1 \cdot 9$$
$$2 \cdot 67 - 9 = 5^3 \cdot 1$$

and we can then write down the first coach:

$$67 \begin{array}{|cccccccc|} 1 & 27 & 8 & 3 & 14 & 24 & 22 & 9 \\ 1 & 1 & 2 & 1 & 1 & 1 & 1 & 3 \\ 0 & 1 & 0 & 0 & 1 & 1 & 1 & 1 \\ 2 & 1 & 1 & 1 & 2 & 2 & 1 & 2 \end{array}_5 .$$

In a similar way we obtain, for the second and third coaches (always beginning the coach with the smallest a_i available):

$$67 \begin{vmatrix} 2 & 13 & 16 & 6 & 28 & 19 & 23 & 18 & 17 & 4 & 26 & 32 & 7 & 12 & 11 & 29 & 21 & 31 & 33 \\ 1 & 1 & 2 & 1 & 1 & 1 & 1 & 1 & 2 & 1 & 1 & 1 & 1 & 1 & 1 & 1 & 1 & 1 & 2 \\ 1 & 0 & 0 & 0 & 0 & 1 & 0 & 0 & 1 & 1 & 0 & 1 & 1 & 1 & 0 & 1 & 0 & 0 & 0 \\ 1 & 1 & 2 & 2 & 1 & 2 & 1 & 1 & 1 & 2 & 2 & 1 & 1 & 1 & 2 & 2 & 2 & 2 & 1 \end{vmatrix}_5 ,$$

$k = 11, \quad c = 3, \quad$ and E is always an *odd* number so that $(-1)^E = -1$.

These three coaches comprise $\Sigma_5(67)$ (but there isn't space to write the complete symbol across the page here!).

From any coach of $\Sigma_5(67)$ we see that the quasi-order of 5 mod 67 is 11 and, in fact,

$$5^{11} \equiv -1 \text{ mod } 67.$$

Once again the coach theorem seems to hold; that is $\Phi(67) = 66 = 2 \cdot 11 \cdot 3$.

*17.4 The generalized coach theorem[4]

The evidence of the last section strongly suggests that an appropriate variation of the coach theorem holds for all t.

It is the purpose of this section to enunciate and prove an important theorem of this chapter. We should warn you that it requires some advanced knowledge of number theory and we wouldn't blame you if you didn't want to absorb all the details before going on. In fact, it isn't necessary to know the proof to appreciate the remaining sections of this chapter.

The generalized coach theorem *Let Φ denote the Euler totient function. Then $\Phi(b) = 2ck$, where $k = \sum_{i=1}^{r} k_i$ and c is the number of coaches (as given in Section 17.2).*

Proof. Let $\mathbb{Z}_b^*$ be the multiplicative group of residues mod b prime to b, so that Order $(\mathbb{Z}_b^*) = \Phi(b)$, and let T be the subgroup of $\mathbb{Z}_b^*$ generated by -1 and 2.

We first observe that, with $t = 2$, where t is the base,

$$T \equiv \{(-1)^i 2^j \text{ mod } b; \quad 0 \le i \le 1, \quad 0 \le j \le k - 1\}, \tag{17.10}$$

with k being the quasi-order of 2 mod b. We now modify (17.10) to accommodate the replacement of base 2 by a general base t; we will then write T_t in place of T, so that $T = T_2$.

---

[4] The preliminary ideas leading up to this section appeared in [23, 24, 47].

Thus we start by modifying (17.10) to handle the case of base t. Then (17.10) is replaced by

$$T_t = \{(-1)^i t^j \bmod b; \quad 0 \le i \le 1, \quad 0 \le j \le k-1\},$$

where b is prime to t, k is the quasi-order of t mod b, $t^k \equiv \pm 1 \bmod b$. Note that, independently of the choice of t,

$$|T_t| = 2k,$$

Thus a **coach** is a symbol as defined in Example 17.1. We will construct the **modified φ-symbol**, and show how the generalized ψ-symbol is related to the generalized φ-symbol. In fact, we will obtain a modified symbol from a given element of $\mathbb{Z}_b^*/T_t$; but it is then an automatic step – which we will describe – to obtain the symbol itself, that is, the coach.

The argument now proceeds in a way similar to the case $t = 2$, as described in [46, 47]. First we show that each element of $\mathbb{Z}_b^*/T_t$ is represented by a number a which is (i) $\not\equiv 0 \bmod t$, (ii) prime to b, and (iii) less than $\frac{b}{2}$. We claim that it is obvious from the structure of T_t that we may represent an element of $\mathbb{Z}_b^*/T_t$ by a number a' which is (i) $\not\equiv 0 \bmod t$, (ii) prime to b, and (iii)' less than b.

Thus let a' be such a number. If $a' < \frac{b}{2}$, there is nothing further to do. But if $a' > \frac{b}{2}$, then, for some q_i, ϵ_i, and t, $\dfrac{q_i b - t^{k_i} a'}{-(-1)^{\epsilon_i}} < \dfrac{b}{2}$ and represents the same element ξ of $\mathbb{Z}_b^*/T_t$. However, $b - a'$ is a multiple of t, so we may set $b - a' = t^\ell a$, with $a \not\equiv 0 \bmod t$, and $\ell \ge 1$. Again a represents the same element of $\mathbb{Z}_b^*/T_t$ as a', and a satisfies conditions (i), (ii), (iii).

We now apply the generalized reverse algorithm φ (that is, the reverse of ψ), given by

$$\frac{q_i b - t^{k_i} a_{i+1}}{-(-1)^{\epsilon_i}} = a_i. \tag{17.11}$$

Thus, writing a_1 for a, we obtain a sequence of numbers $\not\equiv 0 \bmod t$

$$a_1, a_2, \ldots, a_r, a_{r+1}, \tag{17.12}$$

all satisfying conditions (i), (ii), (iii), where

$$a_{r+1} = a_1. \tag{17.13}$$

Explicitly, the passage from a_1 to a_2 is achieved by repeatedly multiplying a_1 by t until we achieve $t^{\ell_1} a_1 > \frac{b}{2}$, with $\ell_1 \ge 1$ and then set $a_2 = \dfrac{q_1 b - t^{\ell_1} a_1}{-(-1)^{\epsilon_1}}$; and we continue, in this way, to generate the entire sequence $a_1, a_2, \ldots, a_r$. Plainly

$a_1, a_2, \ldots, a_r$ all represent the same element of $\mathbb{Z}_b^*/T_t$. Now let us insert the multiples of t, $ta_1, t^2a_1, \ldots, t^{\ell_1}a_1$ between a_1 and a_2, and proceed similarly between a_2 and $a_3, \ldots, a_{r-1}$ and a_r, a_r and a_1. The result is now precisely the effect of the generalized φ-algorithm on the modified symbol. The generalized φ-algorithm is, of course, inverse to the generalized ψ-algorithm used in the construction of the symbol.

Let us now pause to sum up the steps so far. Starting with a_1 satisfying (i), (ii), and (iii), and representing a given element ξ of $\mathbb{Z}_b^*/T_t$, we then apply the generalized (*reverse*) φ-algorithm (17.11) to $a = a_1$, and iterate the applications, obtaining (17.12), namely,

$$a_1, a_2, \ldots, a_r, a_{r+1},$$

all satisfying conditions (i), (ii), (iii) and with $a_{r+1} = a_1$. Thus, starting with a_1 satisfying conditions (i), (ii), (iii) and representing a given element ξ of $\mathbb{Z}_b^*/T$, we apply to a_1 the reverse φ-algorithm, that is, we construct the sequence

$$a_1, ta_1, \ldots, t^{\ell_1}a_1,$$

such that ℓ_1 is minimal for the property $t^{\ell_1}a_1 > \frac{b}{2}$ (so that $\ell_1 \geq 1$), and set

$$a_2 = \frac{q_1 b - t^{\ell_1}a_1}{-(-1)^{\epsilon_1}}.$$

We begin again with a_2 and again apply the generalized φ-algorithm to obtain a_3. We continue in this way until we reach $\ldots, a_r, a_{r+1}$, with $a_{r+1} = a_1$. This must occur eventually since φ is a permutation, indeed, the permutation inverse to the ψ-*algorithm* ([24], p. 116), which was used to construct the **symbol** in the first place.

Indeed, it is easy to see how the symbol may be derived from the element of $\mathbb{Z}_b^*/T_t$ represented by the number a_1 satisfying conditions (i), (ii), (iii). We write down the sequence arising from the process of multiplying by t, together with (17.10), to pass from a_1 to a_2, and then proceed as described above, obtaining

$$a_1, ta_1, \ldots, t^{\ell_1}a_1, a_2, ta_2, \ldots, t^{\ell_2}a_2, a_3, \ldots, a_r, ta_r, \ldots, t^{\ell_r}a_r, a_1 \quad (17.14)$$

It is then not difficult to see that we obtain from (17.14) a **coach** as follows. We write down the terms $\not\equiv 0 \mod t$ in (17.14) as the *modified* top line of the modified coach, thus

$$a_1, a_2, \ldots, a_r, a_1; \quad (17.15)$$

and the modified second line of the coach simply lists the number of t-multiples between successive entries (17.15) in the sequence (17.14). Thus the second line is

$$\ell_1, \ell_2, \ldots, \ell_r. \tag{17.16}$$

The result is a **modified symbol,** from which the true symbol (or *coach*) is obtained by omitting the repeated a_1 from the *start* of (17.15) and then writing each of (17.15), (17.16) backwards.

We demonstrate the procedure with the particular, but not special, case $t = 5$, using the first coach of $\Sigma_5(67)$. The *modified symbol,* incorporating the facts of (17.11), using (17.15), may be written as

$$67 \begin{pmatrix} 1 & 9 & 22 & 24 & 14 & 3 & 8 & 27 & 1 \\ 3 & 1 & 1 & 1 & 1 & 2 & 1 & 1 & \\ 1 & 1 & 1 & 1 & 0 & 0 & 1 & 0 & \\ 2 & 1 & 2 & 2 & 1 & 1 & 1 & 2 & \end{pmatrix}_5 . \tag{17.17}$$

Thus if we start with the element of $\mathbb{Z}_{67}^*/T_5$ represented by the number 1, the sequence (17.14) is given by

$$\frac{2 \cdot 67 - 5^3 \cdot 1}{-(-1)^1} \quad \frac{1 \cdot 67 - 5^1 \cdot 9}{-(-1)^1} \quad \frac{2 \cdot 67 - 5^1 \cdot 22}{-(-1)^1} \quad \frac{2 \cdot 67 - 5^1 \cdot 24}{-(-1)^1} \quad \frac{1 \cdot 67 - 5^1 \cdot 14}{-(-1)^0} \quad \frac{1 \cdot 67 - 5^2 \cdot 3}{-(-1)^0}$$

$$\downarrow \qquad \downarrow \qquad \downarrow \qquad \downarrow \qquad \downarrow \qquad \downarrow$$

$$\underline{1}, 5, 25, 125, \underline{9} \quad , 45, \underline{22} \quad , 110, \underline{24} \quad , 120, \underline{14} \quad , 70, \underline{3} \quad , 15, 75, \underline{8} \quad ,$$

$$\frac{1 \cdot 67 - 5^1 8}{-(-1)^1} \quad \frac{2 \cdot 67 - 5^1 27}{-(-1)^0}$$

$$\downarrow \qquad \downarrow$$

$$40, \underline{27} \quad , \quad \underline{1} \quad .$$

From this sequence we can construct the modified symbol (17.17). Then, by writing (17.17) in reverse, and omitting the repeated "1" we can recover the first coach of $\Sigma_5(67)$. Of course, each of the numbers $1, 27, 8, 3, 14, 24, 22, 9$ represents the same element ξ of $\mathbb{Z}_{67}^*/T_5$.

It is now plain that, in general, the process thus far described sets up a one-to-one correspondence between the set of coaches and the elements of $\mathbb{Z}_b^*/T_t$.

We are now in a position to complete the proof of the generalized coach theorem. We simply have to count the elements in $\mathbb{Z}_b^*/T_t$. But, of course,

$$|\mathbb{Z}_b^*| = \Phi(b) \qquad \text{and} \qquad |T_t| = 2k,$$

as follows easily from (17.17). Thus

$$c = |\mathbb{Z}_b^*|/|T_t| = \Phi(b)/2k,$$

so that

$$\Phi(b) = 2ck. \qquad (17.18)$$

□

Of course, the general coach theorem implies that $\Phi(b)$ is even. However, the standard algorithm for calculating $\Phi(b)$ shows that $\Phi(b)$ is even if b has an odd prime factor, but in general all we have assumed is that b is relatively prime to the base t and $b > 1$. However, it is striking that

$$k \mid \tfrac{1}{2}\Phi(b). \qquad (17.19)$$

This follows easily if b is prime, but is not so obvious if b is composite. However, (17.19) is a trivial consequence of (17.18).

Let us say that b has a **cyclic coach** in base t if $c = 1$; that is, if we can obtain only one coach from b. We then have

The cyclic coach theorem *b has a cyclic coach in base t if, and only if, $\Phi(b) = 2k$, where k is the quasi-order of t mod b.*

An example is given by $\Sigma_2(21)$. In this case $b = 21$, and $\Phi(21) = 12$, $k = 6$. Other examples are

$$\Sigma_3(70), \quad \text{where } \Phi(70) = 24 \quad \text{and} \quad k = 12;$$
$$\Sigma_4(27), \quad \text{where } \Phi(27) = 18 \quad \text{and} \quad k = 9;$$
$$\Sigma_{10}(43), \quad \text{where } \Phi(43) = 42 \quad \text{and} \quad k = 21.$$

17.5 Parlor tricks

The students, Victor and Linda, in the summer research project already knew the content of Chapter 7 and the first four sections of this chapter. They planned to look at many symbols, in order to find interesting patterns among the numbers appearing in those symbols – or to see if there were other interesting questions that could be asked about them. The students had a strong conviction that, if they could spot some patterns, it would lead to interesting results since they, and their instructors, had seen many cases that indicated to them that "In mathematics there are no accidents!" (See [45] for another example of this phenomenon.) At the beginning of their study Linda and Victor used Walden's program to generate a large number of symbols $\Sigma_t(b)$. Having found one particularly interesting pattern, the students devised a rather spectacular number trick, and then supplied the explanation for the trick that appears below.

The trick involves a complete symbol in the case where $t = 2$. The student asks you to look at one of your complete symbols, in base 2, and to tell you the value of b, the largest k_i that appears, and its corresponding a_i. With just this information the student can then line up, in ascending order, all the possible values of a_i with the corresponding k_i underneath them – without calculating $\Sigma(b)$!

Example 17.4 You give the student the information that $b = 35$, that the largest $k_i = 5$, and that the corresponding $a_i = 3$.

The student, after a fairly short calculation that does not involve actually calculating $\Sigma(35)$, writes on the board the following:

$$
\begin{array}{lcccccc}
a_i \rightarrow & 1 & 3 & 9 & 11 & 13 & 17 \\
k_i \rightarrow & 1 & 5 & 1 & 3 & 1 & 1
\end{array}
\tag{17.20}
$$

Before we show you the secret you should try doing this with values obtained either from some of the symbols that have already appeared in this chapter, or with some you have created. Of course, like most magic tricks, once you see how it works it seems trivial.

You may get a hint as to what is going on by first looking at what happens when you are given the data from a symbol for a *prime* number, say 43. In this case the student is told that $b = 43$, the largest $k_i = 5$, and it goes with $a_i = 11$. The student immediately writes on the board the top row of the following array; then, in the second row, she writes 5 under the 11, then in every second space, beginning next to the 5 she writes a 1, then in every fourth place, beginning at the nearest vacant space to 5, she writes a 2, and finally, in every eighth place, beginning at the nearest vacant space to 5, she writes a 3. You can verify, from the symbol for 43, that this result is correct. Do you see why?

$$
\begin{array}{lccccccccccc}
a_i \rightarrow & 1 & 3 & 5 & 7 & 9 & 11 & 13 & 15 & 17 & 19 & 21 \\
k_i \rightarrow & 1 & 3 & 1 & 2 & 1 & 5 & 1 & 2 & 1 & 3 & 1
\end{array}
\tag{17.21}
$$

The following diagram may help to explain the sequence of numbers in the bottom row of array (17.21).

$$
\begin{array}{llccccccccccc}
\text{even numbers } n, & \rightarrow & 42 & 40 & 38 & 36 & 34 & 32 & 30 & 28 & 26 & 24 & 22 \\
42 \geq n \geq 22 & & & & & & & & & & & & \\
\text{greatest power of} & \rightarrow & 1 & 3 & 1 & 2 & 1 & 5 & 1 & 2 & 1 & 3 & 1 \\
\text{2 in } n & & & & & & & & & & & &
\end{array}
\tag{17.22}
$$

Notice that the top row of array (17.22) is simply $43 - a_i$, for $a_i = 1, 3, 5, \ldots, 21$; and the bottom row is the k_i in the equation $43 - a_i = 2^{k_i} a_{i+1}$. To see that this must always happen, draw a number line and write above every even number the highest power of 2 in that number. Note the pattern. In our example, with $b = 43$, we are just looking at the portion of the sequence of the powers of 2 attached to consecutive even numbers between 22 and 42 in reverse order.

So, how does one use this information to obtain the result shown in array (17.21) for $\Sigma(35)$? First, write out all the odd numbers a_i, $1 \le a_i \le 17$, along the top row. Then, *use* the information that the largest $k_i(= 5)$ goes with $a_i = 3$, and fill out the k_i just as in the case $b = 43$ to obtain the array

$$
\begin{array}{lccccccccc}
a_i \rightarrow & 1 & 3 & 5 & 7 & 9 & 11 & 13 & 15 & 17 \\
k_i \rightarrow & 1 & 5 & 1 & 2 & 1 & 3 & 1 & 2 & 7
\end{array}
\tag{17.23}
$$

Finally, use the fact that 35 has 5 and 7 as prime factors – so we really had no right to use any a_i if it has a factor of either 5 or 7 when we produced the original symbol. Thus we now simply *knock out* the entries in the top row of display (17.23) where $a_i = 5, 7$, and 15. This leaves the array shown in (17.20). Now we see why the student needed to do a little scratch work before writing down the array in (17.20). We notice, too, that the argument shows that there will always be just one largest k_i, because there is always just one suitable odd number for which $b - a_i = 2^{k_i} a_{i+1}$.

It now seemed natural for the students to ask: Can an analogous trick be performed for any symbol $\Sigma_t(b)$? That is, given the values of b and t, along with the value of the largest k_i and the corresponding a_i, could the student line up all the possible a_i, in ascending order, and tell the value of the corresponding k_i, without constructing the complete symbol?

It is not at all surprising that in the general case you must be told information not only about the largest k_i, but also about the corresponding ϵ_i and q_i in order to be able to determine all the corresponding values in the column under the rest of the a_is. It is also not surprising that this situation is much more complex – probably too complex to make a parlor trick out of it, especially for large values of t.

To show you just how the analysis goes let us refer to a particular but not special case, namely the complete symbol $\Sigma_5(67)$, which was calculated in Section 17.3. The first coach of the complete symbol $\Sigma_5(67)$ is given by

$$
67 \left|
\begin{array}{cccccccc}
1 & 27 & 8 & 3 & 14 & 24 & 22 & 9 \\
1 & 1 & 2 & 1 & 1 & 1 & 1 & 3 \\
0 & 1 & 0 & 0 & 1 & 1 & 1 & 1 \\
2 & 1 & 1 & 1 & 2 & 2 & 1 & 2
\end{array}
\right|_5 ,
$$

Table 17.1 *When $a_i \equiv 2 \bmod 5$.*

a_i	32	27	22	17	12	7	2
k_i	1	1	1	2	1	1	1
ϵ_i	1	1	1	1	1	1	1
q_i	1	1	1	1	1	1	1

Table 17.2 *When $a_i \equiv 3 \bmod 5$.*

$a_i \rightarrow$	3	8	13	18	23	28	33
$k_i \rightarrow$	1	2	1	1	1	1	2
$\epsilon_i \rightarrow$	0	0	0	0	0	0	0
$q_i \rightarrow$	1	1	1	1	1	1	1

Table 17.3.

multiples, m, of 5, $35 \leq m \leq 100$ $\rightarrow$	35	40	45	50	55	60	65	$\boxed{67}$	70	75	80	85	90	95	100
greatest power of 5 in m $\rightarrow$	1	1	1	2	1	1	1	$\uparrow$	1	2	1	1	1	1	2

and the second and third coaches of $\Sigma_5(67)$, always beginning the coach with the smallest available a_i, may be displayed as

$$67 \begin{vmatrix} 2 & 13 & 16 & 6 & 28 & 19 & 23 & 18 & 17 & 4 & 26 & 32 & 7 & 12 & 11 & 29 & 21 & 31 & 33 \\ 1 & 1 & 2 & 1 & 1 & 1 & 1 & 1 & 2 & 1 & 1 & 1 & 1 & 1 & 1 & 1 & 1 & 1 & 2 \\ 1 & 0 & 0 & 0 & 0 & 1 & 0 & 0 & 1 & 1 & 0 & 1 & 1 & 1 & 0 & 1 & 0 & 0 & 0 \\ 1 & 1 & 2 & 2 & 1 & 2 & 1 & 1 & 1 & 2 & 2 & 1 & 1 & 1 & 2 & 2 & 2 & 2 & 1 \end{vmatrix}_5.$$

From this we readily see that

$$k = 11, \quad c = 3, \quad \text{and, since } E \text{ is always an } odd \text{ number, } (-1)^E = -1.$$

It turns out to be instructive to list the values for a_i, k_i, ϵ_i, and q_i in 4 subsets where $a_i \equiv n \bmod 5$, with $n = 1, 2, 3$ or 4. It is also useful to list the multiples of 5 that one obtains when using

$$q_i b + (-1)^{\epsilon_i} a_i = 5^{k_i} a_{i+1}, \quad i = 1, 2, 3, 4 \quad (a_5 = a_1),$$

to create the symbol.

Now, compare the entries in the complete symbol $\Sigma_5(67)$ with the data given in Tables 17.1–17.6, looking for relevant patterns. In order to help you to *see* the patterns more clearly we have written the a_is in ascending order when a_i is congruent to an odd number mod 5 (in Tables 17.2 and 17.5), and in descending order when a_i is congruent to an even number mod 5 (in Tables 17.1 and 17.4).

The bottom row of Table 17.3 shows, on the left of 67, the quantities $67 - a_i$, where (reading from left to right) $a_i = 32, 27, 22, 17, 12, 7$, and 2, respectively. Similarly the bottom row of Table 17.3 shows, on the right of 67, the quantities $67 + a_i$ where (reading from left to right) $a_i = 3, 8, 13, 18, 23, 28$, and 33, respectively. Compare the second row of Table 17.3 with the second rows of Table 17.1 and Table 17.2 (reading from left to right).

Table 17.4 *When $a_i \equiv 4 \bmod 5$.*

$a_i \rightarrow$	29	24	19	14	9	4
$k_i \rightarrow$	1	1	1	1	3	1
$\epsilon_i \rightarrow$	1	1	1	1	1	1
$q_i \rightarrow$	2	2	2	2	2	2

Table 17.5 *When $a_i \equiv 1 \bmod 5$.*

$a_i \rightarrow$	1	6	11	16	21	26	31
$k_i \rightarrow$	1	1	1	2	1	1	1
$\epsilon_i \rightarrow$	0	0	0	0	0	0	0
$q_i \rightarrow$	2	2	2	2	2	2	2

Table 17.6.

multiples, m, of 5, $105 \leq m \leq 165$ $\rightarrow$	105	110	115	120	125	130	134	135	140	145	150	155	160	165
greatest power of 5 in m $\rightarrow$	1	1	1	1	3	1	$\uparrow$	1	1	1	2	1	1	1

The bottom row of Table 17.6 shows, on the left of 134, the quantities $2 \cdot 67 - a_i$, where (reading from left to right) $a_i = 29, 24, 19, 14, 9$, and 4, respectively. Similarly the bottom row of Table 17.6 shows, on the right of 134, the quantities $2 \cdot 67 + a_i$, where (reading from left to right) $a_i = 1, 6, 11, 16, 21, 26$, and 31, respectively. Compare the second row of Table 17.6 with the second rows of Table 17.4 and Table 17.5 (reading from left to right).

Perhaps what stands out first is that, for each subset, the values of ϵ_i and q_i are the same for all a_i in that set. This is, of course, no accident! As we have said "in mathematics there are no accidents." What this means is that, with k_i always maximal, then if

$$
\begin{cases}
\epsilon_i = q_i = 1, & \text{the values of } k_i \text{ must satisfy} \quad 67 - a_i = 5^{k_i} a_{i+1}, \\
\epsilon_i = 0, \quad q_i = 1, & \text{the values of } k_i \text{ must satisfy} \quad 67 + a_i = 5^{k_i} a_{i+1}, \\
\epsilon_i = 1, \quad q_i = 2, & \text{the values of } k_i \text{ must satisfy} \quad 2 \cdot 67 - a_i = 5^{k_i} a_{i+1}, \\
\epsilon_0 = 0, \quad q_i = 2, & \text{the values of } k_i \text{ must satisfy} \quad 2 \cdot 67 + a_i = 5^{k_i} a_{i+1}.
\end{cases}
$$

This explains what is happening and the reader can well believe that it is no trivial matter to do the "parlor trick" for large t. Still, we have seen that there are lovely patterns among these numbers, and we now understand some of the features, and inner workings, of the complete symbol better than we did before.

17.6 A little linear algebra

In [42] it was shown that, in the case when $t = 2$, it is possible to make up *any* folding sequence and then to calculate what regular star polygons could be constructed from this folded paper. A brief explanation of this result was given in Section 7.5.

The students wanted to see if the analogous result could be obtained for any base t. Of course, once they saw the system of linear equations that were generated in the more general case they realized that it was simply a matter of carrying out the procedure outlined in Chapter 7, with the appropriate variations. The interesting part of the study was observing how wonderfully symmetric – indeed, how beautiful – the solution turned out to be.

Since the techniques concerning linear algebra are already described in [42], and because there is now software available that will readily solve these systems of linear equations, the students focused their attention on writing the solution of some special cases in such a way as to be able to see how to write the general solution for any base t, given values for the k_i, ϵ_i, and q_i that made sense.

We first demonstrate what happens with a couple of examples for particular, but not special, cases. Then we give the solution which, we believe, encapsulates all that one needs to know in order to find the suitable a_1 and b for any given situation.

Example 17.5 Suppose the values for k_i, ϵ_i, and q_i are known in the symbol

$$b \begin{vmatrix} a_i & a_2 & a_3 \\ k_1 & k_2 & k_3 \\ \epsilon_1 & \epsilon_2 & \epsilon_3 \\ q_1 & q_2 & q_3 \end{vmatrix}_4 . \tag{17.24}$$

Is it possible to recover the values for b and a_1 (and, hence, also for a_2 and a_3)? Since $t = 4$, we know that $q_i = 1$ or 2 and that

$$\text{when} \quad q_i = 1, \quad \epsilon_i \text{ can be 1 or 0,}$$

and

$$\text{when} \quad q_i = 2, \quad \epsilon_i \text{ must be 1.}$$

We also know that $\sum k_i = k$ and that $r = 3$.
Using the formula (17.8), with $t = 4$, we have

$$q_i b + (-1)^{\epsilon_i} a_i = 4^{k_i} a_{i+1}, \quad i = 1, 2, 3 \quad (a_4 = a_1),$$

from which we obtain the following system of 3 linear equations:

$$q_1 b + (-1)^{\epsilon_1} a_1 = 4^{k_1} a_2$$
$$q_2 b + (-1)^{\epsilon_2} a_2 = 4^{k_2} a_3$$
$$q_3 b + (-1)^{\epsilon_3} a_3 = 4^{k_3} a_1.$$

Writing this system of equations in a more suggestive form we have:

$$
\begin{aligned}
-(-1)^{\epsilon_1}a_1 + 4^{k_1}a_2 \quad &+ 0 &&= q_1 b \\
0 \quad - (-1)^{\epsilon_2}a_2 &+ 4^{k_2}a_3 &&= q_2 b \\
4^{k_3}a_1 \quad + 0 \quad &- (-1)^{\epsilon_3}a_3 &&= q_3 b.
\end{aligned}
$$

To find a_1 we proceed, using Cramer's rule for determinants, to obtain a_1. Let the determinant, consisting of the coefficients of a_1, a_2, a_3 on the left-hand side, be denoted by B and let the numerator, where the coefficients of a_1 are replaced by $q_1 b$, $q_2 b$, $q_3 b$, be denoted A_1. Thus

$$
B = \begin{vmatrix} -(-1)^{\epsilon_1} & 4^{k_1} & 0 \\ 0 & -(-1)^{\epsilon_2} & 4^{k_2} \\ 4^{k_3} & 0 & -(-1)^{\epsilon_3} \end{vmatrix}, \qquad A_1 = b\begin{vmatrix} q_1 & 4^{k_1} & 0 \\ q_2 & -(-1)^{\epsilon_2} & 4^{k_2} \\ q_3 & 0 & -(-1)^{\epsilon_3} \end{vmatrix}.
$$

Hence, the solution for a_1 is given by

$$
a_1 = \frac{A_1}{B} = \frac{b\begin{vmatrix} q_1 & 4^{k_1} & 0 \\ q_2 & -(-1)^{\epsilon_2} & 4^{k_2} \\ q_3 & 0 & -(-1)^{\epsilon_3} \end{vmatrix}}{\begin{vmatrix} -(-1)^{\epsilon_1} & 4^{k_1} & 0 \\ 0 & -(-1)^{\epsilon_2} & 4^{k_2} \\ 4^{k_3} & 0 & -(-1)^{\epsilon_3} \end{vmatrix}}
$$

$$
= \frac{b[q_1(-1)^{\epsilon_2 + \epsilon_3} + q_3 4^{k_1 + k_2} + q_2(-1)^{\epsilon_3}4^{k_1}]}{4^{k_1 + k_2 + k_3} - (-1)^{\epsilon_1 + \epsilon_2 + \epsilon_3}}. \tag{17.25}
$$

Using the fact that we have $r = 3, k = \sum k_i, E = \sum \epsilon_i$, and knowing what happens subsequently, we rewrite (17.25) in the equivalent form

$$
\frac{a_1}{b} = \frac{q_3 4^{k_1 + k_2} + q_2(-1)^{\epsilon_3}4^{k_1} + q_1(-1)^{\epsilon_2 + \epsilon_3}}{4^k + (-1)^3(-1)^E}. \tag{17.26}
$$

The reason for expressing our answer in this form is 2-fold. First, in our symbol we must have $\gcd(a_1, b) = 1$. Thus in order to make sure we have the desired values for a_1 and b we must reduce the fraction $\frac{a_1}{b}$ to its lowest terms. The second reason is that the right-hand side is more suggestive of what we will obtain in the more complicated situation.

Let us do a reality check. Suppose we have the symbol

$$
b\begin{vmatrix} a_1 & a_2 & a_3 \\ 1 & 4 & 2 \\ 0 & 1 & 1 \\ 1 & 2 & 1 \end{vmatrix}_4 .
$$

Using the solution given in (17.26) we have

$$\frac{a_1}{b} = \frac{(1)4^5 + 2(-1)4^1 + (1)(-1)^2}{4^7 + (-)^3(-1)^2} = \frac{1017}{16\,383} \text{ (reduced)} = \frac{339}{5461}.$$

Since $\gcd(339, 5461) = 1$ we know that $b = 5461$ and $a_1 = 339$. That this is true is easy to check, yielding the following coach from $\Sigma_4(5461)$.

$$5461 \begin{vmatrix} 339 & 1450 & 37 \\ 1 & 4 & 2 \\ 0 & 1 & 1 \\ 1 & 2 & 1 \end{vmatrix}_4 ,$$

which, according to our convention of beginning each coach with the smallest a_i available, would appear in $\Sigma_4(5461)$ as

$$5461 \begin{vmatrix} 37 & 339 & 1450 \\ 2 & 1 & 4 \\ 1 & 0 & 1 \\ 1 & 1 & 2 \end{vmatrix}_4 .$$

Suppose some trouble-maker gives you data inconsistent with the conditions of the generalized quasi-order theorem. For example, suppose he gives you the incomplete symbol

$$b \begin{vmatrix} a_i & a_2 & a_3 \\ 1 & 2 & 3 \\ 0 & 1 & 1 \\ 2 & 1 & 2 \end{vmatrix}_4 \tag{17.27}$$

and asks you to complete it. You should say, immediately, "this is impossible, because for $t = 4$, if $q_i = 2$ then the corresponding ϵ_i *must* be 1." The first column in this incomplete symbol violates this condition. However, if you don't notice this and work out the answer for a_1, using (17.26) you will obtain

$$a_1 = \frac{(2)4^3 + (1)(-1)^14 + (2)(-1)^2}{4^6 + (-1)^3(-1)^2} = \frac{126}{4095} \text{ (reduced)} = \frac{2}{65}.$$

Then if you construct the first coach of the symbol with $b = 65$ and $a_1 = 2$ you will obtain

$$65 \begin{vmatrix} 2 \\ 3 \\ 1 \\ 2 \end{vmatrix}_4$$

which is obviously not the completion of (17.27). If that isn't enough proof to satisfy you that something has gone wrong, and you try to use

$$q_i b + (-1)^{\epsilon_i} a_i = 4^{k_i} a_{i+1}, \quad i = 1, 2, \ldots, r \quad \text{(where } a_{r+1} = a_1\text{)},$$

with $b = 65$, $a_1 = 2$, $\epsilon_1 = 0$, and $q = 2$, you obtain

$$2 \cdot 65 + (-1)^0 2 = 132 = 4^1 \cdot 33,$$

so that $a_2 = 33$. But this violates the condition that $a_i < \frac{b}{2}$. Thus we see that, if we pay attention, the mathematics keeps us honest. If we persist long enough in the wrong direction good mathematics will always try to tell us that something is amiss!

We now present the **big guess** that puts us so close to the general formula that an observant reader might feel nothing more is necessary. However, we want to point out that this is just an example of a special, but not particular case, that encapsulates all of the features of the general case. We will let the ambitious reader work out the statement of the general case.

On the basis of the formulas the students obtained for $t = 2, 3$, and 4 they made a guess for what the formula for recovering b and a_1 from the following coach would look like

$$b \begin{vmatrix} a_1 & a_2 & a_3 & a_4 & a_5 & a_6 & a_7 & a_8 \\ k_1 & k_2 & k_3 & k_4 & k_5 & k_6 & k_7 & k_8 \\ \epsilon_1 & \epsilon_2 & \epsilon_3 & \epsilon_4 & \epsilon_5 & \epsilon_6 & \epsilon_7 & \epsilon_8 \\ q_1 & q_2 & q_3 & q_4 & q_5 & q_6 & q_7 & q_8 \end{vmatrix}_5. \tag{17.28}$$

In this case we have $r = 8$, $k = \sum k_i$, $E = \sum \epsilon_i$, and

$$\text{if } \begin{cases} \epsilon_i = 1, & q_i \text{ can be 1 or 2} \\ \epsilon_i = 0, & q_i \text{ can be 1 or 2.} \end{cases}$$

The guess is that

$$\frac{a_1}{b} = \left[q_8 5^{\sum_{k=1}^{7} k_i} + q_7 (-1)^{\sum_{i=8}^{8} \epsilon_i} 5^{\sum_{k=1}^{6} k_i} + q_6 (-1)^{\sum_{i=7}^{8} \epsilon_i} 5^{\sum_{k=1}^{5} k_i} \right.$$

$$+ q_5 (-1)^{\sum_{i=6}^{8} \epsilon_i} 5^{\sum_{k=1}^{4} k_i} + q_4 (-1)^{\sum_{i=5}^{8} \epsilon_i} 5^{\sum_{k=1}^{3} k_i} + q_3 (-1)^{\sum_{i=4}^{8} \epsilon_i} 5^{\sum_{k=1}^{2} k_i}$$

$$\left. + q_2 (-1)^{\sum_{i=3}^{8} \epsilon_i} 5^{\sum_{k=1}^{1} k_i} + q_1 (-1)^{\sum_{i=2}^{8} \epsilon_i} \right] \Big/ \left[5^k - (-1)^r (-1)^E \right]. \tag{17.29}$$

Checking this, in a particular but not special case, we suppose we are given the following incomplete coach

$$
b\begin{vmatrix} a_1 & a_2 & a_3 & a_4 & a_5 & a_6 & a_7 & a_8 \\ 2 & 1 & 1 & 1 & 1 & 1 & 1 & 1 \\ 0 & 0 & 1 & 1 & 0 & 1 & 0 & 1 \\ 1 & 1 & 1 & 2 & 2 & 1 & 2 & 2 \end{vmatrix}_5 , \quad \text{with } k = 9, \quad r = 9. \tag{17.30}
$$

Using (17.29) we obtain

$$
\frac{a_1}{b} = [2 \cdot 5^8 + 2 \cdot (-1)^1 5^7 + 1 \cdot (-1)^1 5^6 + 2 \cdot (-1)^2 5^5 + 2 \cdot (-1)^2 5^4
$$

$$
+ 1 \cdot (-1)^3 5^3 + 1 \cdot (-1)^4 5^2 + 1 \cdot (-1)^4] / [5^9 - (-1)^8 (-1)^4]
$$

$$
= \frac{781\,250 - 156\,250 - 15\,625 + 6250 + 1250 - 125 + 25 + 1}{1953\,124}
$$

$$
= \frac{616\,776}{1953\,124}; \quad \text{then, dividing top and bottom by } 102\,796 \text{ this}
$$

$$
= \frac{6}{19}.
$$

Thus we see that $b = 19$, and $a_1 = 6$. Using the generalized quasi-order theorem for these values of b and a_i, we calculate, in order for each a_i, the numbers

$$
b - a_i, \quad b + a_i, \quad 2b - a_i, \quad \text{and} \quad 2b + a_i,
$$

stopping at the *first* number that has 5 as a factor. These calculations tell us the values for q_i, ϵ_i, and a_{i+1} that we need to record for this coach of the complete symbol.

The calculations readily show that the symbol for this coach takes the form (compare with (17.30))

$$
19\begin{vmatrix} 6 & 1 & 4 & 3 & 7 & 9 & 2 & 8 \\ 2 & 1 & 1 & 1 & 1 & 1 & 1 & 1 \\ 0 & 0 & 1 & 1 & 0 & 1 & 0 & 1 \\ 1 & 1 & 1 & 2 & 2 & 1 & 2 & 2 \end{vmatrix}_5 \quad \text{– a triumph for the method!}
$$

Of course, in accordance with our convention of beginning each coach with the smallest available a_i, this would be written beginning with $a_1 = 1$, though this is not so important in this case since, as you can tell from the values of a_i in the top row, $\Sigma_5(19)$ has only one coach!

17.7 Some open questions

We leave the reader with some open questions.

(1) How can you tell if a complete symbol $\Sigma_t(b)$ will have only one coach? How can you predict how many coaches a symbol will have?

(2) How can you tell what the value of r will be for a given coach? Is it the case that the smallest r always occurs in the first coach (where $a_1 = 1$)?

(3) Given b, how are the values of r for different coaches in the complete symbol related? Can you tell in which coach the largest r will occur?

(4) Given b, can you tell which collection of numbers will come together as the top row of a coach?

(5) For a fixed k, can the sequence of numbers $k_1, k_2, \ldots, k_r$ appear in the bottom row of two distinct coaches? (The answer is no: see page 135 of [23].) Are there simple *a priori* necessary and/or sufficient conditions, beyond $\sum k_i = k$, that can tell you whether such a sequence is a bottom row of a coach, short of running the entire computation?

References

[1] Alexanderson, G. L. and Wetzel, J. E., Arrangements of planes in space, *Discrete Math.* **34** (1981), 219–240.

[2] Ball, W. W. R., *Mathematical Recreations and Essays,* rev. H. S. M. Coxeter, Wildhern Press, Edinburgh (2009).

[3] Beck, A., Bleicher, M. N., and Crowe, D. W., *Excursions into Mathematics*, Worth Publishers, New York (1969).

[4] Berkove, E. J. and Dumont, J. P., It's okay to be square if you're a flexagon, *Math. Mag.* **77**, No. 5 (2004), 335–348.

[5] Chandler, D. L., *The Boston Globe*, Feb. 17, 1988.

[6] Cohn, P. M., *Bull. London Math. Soc.* **43** (2002), 613–618.

[7] Courant, R. and Robbins, H., *What is Mathematics?*, Oxford University Press (1969).

[8] Coxeter, H. S. M., *Regular Polytopes*, second edition, Macmillan, New York (1963).

[9] Coxeter, H. S. M., du Val, P., Flather, H. T., and Petrie, J. F., *The Fifty-Nine Icosahedra*, Springer-Verlag, New York (1982).

[10] Cromwell, P. R., *Polyhedra*, Cambridge University Press (1997).

[11] Cundy, H. M., Antiprism frameworks, *Math. Gaz.* **61**, No. 417 (1977), 182–187.

[12] Cundy, H. M. and Rollett, A. P., *Mathematical Models*, second edition, Oxford University Press, New York (1973).

[13] Descartes, René, *Oeuvres*, vol. X, 265–269.

[14] Euler, Leonhard, *Opera Omnia*, series I, Orell Füssli Verlag, vol. 26 (1953), XIV–XVI, 71–108 and 217–218.

[15] Froemke, J. and Grossman, G. W., An algebraic approach to some number-theoretical problems arising from paper-folding regular polygons, *Amer. Math. Monthly* **95**, No. 4 (1998), 289–307.

[16] Gardner, M., *Scientific American*, January, 1956.

[17] Gardner, M., *The Scientific American Book of Mathematical Puzzles and Diversions*, Simon & Schuster, New York (1959).

[18] Gilpin, M., Symmetries of the trihexaflexagon, *Math. Mag.* **49**, No. 4 (1976), 189–192.

[19] Grünbaum, B. and Shephard, G. C., Satins and twills – an introduction to the geometry of fabrics, *Math. Mag.* **53**, No. 3 (1980), 139–166.

[20] Grünbaum, B. and Shephard, G. C., Isonemal fabrics, *Amer. Math. Monthly* **95**, No. 1 (1988), 5–30.

[21] Grünbaum, B. and Shephard, G. C., The geometry of fabrics, *Geometrical Combinatorics* (Milton Keynes) *Research Notes in Mathematics* **114**, Pitman, Boston (1984), 77–98.

[22] Grünbaum, B. and Shephard, G. C., A dual for Descartes' theorem on polyhedra. *Math. Gaz.* **71** (1987), 214–216.

[23] Hilton, P., Holton, D., and Pedersen, J., *Mathematical Reflections – In a Room with Many Mirrors,* Undergraduate Texts in Mathematics, Springer-Verlag, New York (1997).

[24] Hilton, P., Holton, D., and Pedersen, J., *Mathematical Vistas – In a Room with Many Windows*, Undergraduate Texts in Mathematics, Springer-Verlag, New York (2002).

[25] Hilton, P. and Pedersen, J., Descartes, Euler, Poincaré, Pólya and polyhedra, *L'Enseign. Math.* **27** (1981), 327–343.

[26] Hilton, P. and Pedersen, J., Approximating any regular polygon by folding paper; An interplay of geometry, analysis and number theory, *Math. Mag.* **56** (1983), 141–155.

[27] Hilton, P. and Pedersen, J., Folding regular star polygons and number theory, *Math. Intelligencer* **7**, No. 1 (1983), 15–26.

[28] Hilton, P. and Pedersen, J., On certain algorithms in the practice of geometry and the theory of numbers, *Publ. Sec. Mat. Univ. Autónoma Barcelona* **29**, No. 1 (1985), 31–64.

[29] Hilton, P. and Pedersen, J., On the complementary factor in a new congruence algorithm, *Int. J. Math. and Math. Sci.* **10**, No. 1 (1987), 113–123.

[30] Hilton, P. and Pedersen, J., *Geometry in Practice and Numbers in Theory, Monographs in Undergraduate Mathematics* **16** (1987). Available from Department of Mathematics, Guilford College, Greensboro, NC 27410.

[31] Hilton, P. and Pedersen, J., Discovering, modifying and solving problems: a case study from the contemplation of polyhedra. *Teaching and Learning: A Problem-Solving Focus*, NCTM (1987), 47–91.

[32] Hilton, P. and Pedersen, J., On a generalization of folding numbers, *Southeast Asian Bull. Math.* **12**, No. 1 (1988), 53–63.

[33] Hilton, P. and Pedersen, J., *Build your own Polyhedra*. Addison-Wesley, Menlo Park, CA (1988); (reprinted by Pearson Learning, in 1994).

[34] Hilton, P. and Pedersen, J., Duality and Descartes' angular deficiency, *Computers Math. Appl.* **17**, No. 1–3 (1989), 73–88.

[35] Hilton, P. and Pedersen, J., Geometry: A gateway to understanding, *College Math. J.* **24**, No. 4 (1993), 298–317.

[36] Hilton, P. and Pedersen, J., Some arithmetical connections between folding numbers and generalized Fibonacci and Lucas numbers, *Bulletin de la Société Mathématique de Belgique* (Series B) **45** (3) (1993), 273–296.

[37] Hilton, P. and Pedersen, J., On factoring $2^n \pm 1$, *Math. Educator* **5**, No. 1 (1994), 29–32.

[38] Hilton, P. and Pedersen, J., Connecting geometry and number theory, in *In Eves' Circles*, The Mathematical Association of America/Notes Number 34 (1994), 17–39.

[39] Hilton, P. and Pedersen, J., On folding instructions for products of Fermat numbers, *SEA Bull. Math.* **18**, No. 2 (1994), 19–27.

[40] Hilton, P. and Pedersen, J., The Euler characteristic and Pólya's dream, *Amer. Math. Monthly* **103**, No. 2 (1996), 121–131.

[41] Hilton, P. and Pedersen, J., Symmetry in practice – Recreational constructions, visual mathematics (an electronic publication), http://members.tripod.com/vismath/hil/ Also published in *Symmetry: Culture and Science 8,* Nos. 3–4 (1997), 409–429.

[42] Hilton, P. and Pedersen, J., The unity of mathematics: A casebook comprising practical geometry, number theory and linear algebra, *Teaching Mathematics and Computer Science* **1** (2003), 1–34.

[43] Hilton, P. and Pedersen, J., Thoughts on an optimistic expectation of Abbé Mersenne, *Math. Gaz.* **88**, No. 531 (2004), 503–508.

[44] Hilton, P., Pedersen, J., Quintanar-Zilinskas, V., Velarde, L., and Walden, B., Patterns relating to complete symbols: A research project connected with paper-folding and number-theory (to appear in *SEAMS*).

[45] Hilton, P., Pedersen, J., and Ross, P., In mathematics there are no accidents, *Menemui Matematik* **9**, No. 3 (1987), 121–143.

[46] Hilton, P., Pedersen, J., and Walden, B., A property of complete symbols: An ongoing saga connecting geometry and number theory (to appear in *Homage to a Pied Piper*, A. K. Peters, Ltd.)

[47] Hilton, P., Pedersen, J., and Walden, B., Paper-folding, polygons, complete symbols, and the Euler totient function: An ongoing saga connecting geometry, algebra, and number theory *Second International Congress of Algebra and Combinatorics*, *WSCP-Proceedings* (2008).

[48] Hilton, P., Pedersen, J., and Walser, H., The faces of the tri-hexaflexagon, *Math. Mag.* **70**, No. 4 (1997), 243–251 (and the figure on the cover).

[49] Holden, A., *Shapes, Space and Symmetry,* Columbia University Press, New York (1971).

[50] Kerr, J. W. and Wetzel, J. E., Platonic divisions of space, *Math. Mag.* **51**, No. 4 (1978), 229–234.

[51] Lakatos, L., *Proofs and Refutations: The Logic of Mathematical Discovery*, Cambridge University Press (1976).

[52] Ledermann, W. (Chief Editor), *Handbook of Applicable Mathematics, Supplement*, John Wiley, New York (1990).

[53] McFarlane, C. and Withers, W. D., Dynamical systems and irrational angle construction by paper-folding, *Amer. Math. Monthly* **115** (2008), 355–358.

[54] Oakley, C. L. and Wisner, R. J., Flexagons, *Amer. Math. Monthly* **64** (1957), 143–154.

[55] Pedersen, J. J., Asymptotic Euclidean type constructions without Euclidean tools, *Fibonacci Quart.* **9** (1971), 199–216.

[56] Pedersen, J. J., Sneaking up on a group, *Two-Year College Math. J.* **3** (1972), 9–12.

[57] Pedersen, J. J., Collapsoids, *Math. Gaz.* **59** (1975), 81–94.

[58] Pedersen, J. J., Braided rotating rings, *Math. Gaz.* **62** (1978), 15–18.

[59] Pedersen, J. J., Visualising parallel divisions of space, *Math. Gaz.* **62** (1978), 250–262.

[60] Pedersen, J. J., Jennifer's puzzle, *Matimya's Matematika* **4**, No. 4 (1980), 10–18.

[61] Pedersen, J. J., Tetraflexagons, *Math. Digest* (South Africa) **46** (1982), 1–3.

[62] Pedersen, J. J., Some isonemal fabrics on polyhedral surfaces, *The Geometric Vein*: The Coxeter Festschrift, ed. C. Davis, B. Grünbaum and F. A. Sherk, Springer-Verlag, New York (1982), 99–122.

[63] Pedersen, J., Parallel divisions of space, in *Mathematical Adventures for Students and Amateurs*, ed. D. Hayes and T. Shubin, Spectrum Series, (MAA) (2004), 117–133.

[64] Pedersen, J. J. and Pedersen, K. A., *Geometric Playthings to Color, Cut, and Fold*, Dale Seymour Publications, Upper Sadde River (1973).

[65] Pedersen, J. and Pólya, G., On problems with solutions attainable in more than one way, *College Math. J.* **15** (1984), 218–228.

[66] Polster, B., Variations on a theme in paper-folding, *Amer. Math. Monthly* **111** (2004), 39–47.

[67] Pólya, G., *Induction and Analogy in Mathematics,* Vol. I of *Mathematics and Plausible Reasoning*, Princeton University Press (1954).

[68] Pólya, G., Intuitive outline of the solution of a basic combinatorial problem, in *Switching Theory in Space Technology*, ed. H. Siken and W. F. Mann, Stanford University Press (1963), 3–7.

[69] Pólya, G., Guessing and proving, *Two-Year College Math. J.* **9**, No. 1 (1978), 1–9.

[70] Pólya, G., *Mathematical Discovery*, combined edition, John Wiley, New York (1981).

[71] Pook, L., *Flexagons Inside Out*, Cambridge University Press (2000).

[72] Pugh, A., *Polyhedra – a Visual Approach*, University of California Press, Berkeley (1976).

[73] Schäfli, L., Theorie der vielfachen Kontinuität, *Denkschriften der Schweizerischen naturforschenden Gesellschaft* **38** (1901), 1–237.

[74] Schattschneider, D. and Walker, W., *M. C. Escher Kaleidocycles*, Pomegranate Artbooks, Corte Madera, CA. (1987).

[75] *SIAM News* **35**, No. 7, Sept. 6, 2002.

[76] Steiner, J., Einige Gesetze über die Theilung der Ebene und des Räumes, *J. reine Angew. Math.* **1** (1826), 349–364

[77] Walser, H., *Symmetry*, translated by P. Hilton and J. Pedersen, Washington, D. C., MAA (2000).

[78] Walser, H., *The Golden Section*, translated by P. Hilton and J. Pedersen, Washington, D. C., MAA (2001).

[79] Wenninger, M. J., *Polyhedron Models for the Classroom*, National Council of Teachers of Mathematics, Supplementary Publication, Washington, D.C. (1968).

[80] Wenninger, M. J., *Polyhedron Models*, Cambridge University Press, New York (1971).

[81] Weyl, H., *Symmetry*, Princeton University Press (1952).

[82] Wilde, C. O., *The Contraction Mapping Principle*, UMAP Unit 326 (1983) (Lexington, MA: Comap, Inc.).

[83] Willoughby, S. S., Bereiter, C., Hilton, P., and Rubinstein, J. H., *Mathematical Explorations and Applications*, SRA McGraw-Hill, New York (1998).

Index